Industrial Waste Treatment

Industrial Waste Treatment

Maria Malone

SYRAWOOD
PUBLISHING HOUSE
New York

Published by Syrawood Publishing House,
750 Third Avenue, 9th Floor,
New York, NY 10017, USA
www.syrawoodpublishinghouse.com

Industrial Waste Treatment
Maria Malone

International Standard Book Number: 978-1-64740-143-6 (Hardback)

Cataloging-in-Publication Data

Industrial waste treatment / Maria Malone.
p. cm.
Includes bibliographical references and index.
ISBN 978-1-64740-143-6
1. Factory and trade waste. 2. Sewage--Purification. 3. Sewage disposal plants. I. Malone, Maria.
TD897.6 .I53 2022
628.4--dc23

TABLE OF CONTENTS

PREFACE

It is with great pleasure that I present this book. It has been carefully written after numerous discussions with my peers and other practitioners of the field. I would like to take this opportunity to thank my family and friends who have been extremely supporting at every step in my life.

The waste produced by industrial activities during the manufacturing processes in factories, mills and mines is termed as industrial waste. The most common types of industrial waste include masonry and concrete, dirt and gravel, oil solvents and scrap lumber. Industrial waste can be in different forms such as solid, liquid or gaseous. It can be hazardous or non-hazardous waste. Hazardous waste can be reactive, corrosive, ignitable, toxic and radioactive. Industrial waste is designated as chemical waste, toxic waste, municipal solid waste and industrial solid waste. The industrial waste consisting of conventional pollutants such as biochemical oxygen demand can be treated at sewage treatment plants. Industrial waste that contains toxic pollutants is managed through specialized treatment systems. This book outlines the processes and applications of industrial waste management in detail. The topics covered herein deal with the core aspects of this field. This textbook is appropriate for students seeking detailed information in this area as well as for experts.

The chapters below are organized to facilitate a comprehensive understanding of the subject:

Chapter – Industrial Waste and its Types

The waste produced by the industries is referred to as industrial waste. It can be classified into toxic waste, hazardous waste, radioactive waste, household hazardous waste, chemical waste, solid waste, etc. This is an introductory chapter which will briefly introduce all the significant aspects of industrial wastes and its types.

Chapter – Sources of Industrial Waste

Sources of industrial waste include nondurable goods, durable goods, containers packaging, and food and yard waste. Other sources include battery manufacturing, organic chemicals manufacturing, food industry, iron and steel industry, etc. are the sources of industrial wastewater. The topics elaborated in this chapter will help in gaining a better perspective about these sources.

Chapter – Impacts of Industrial Waste

Industrial waste impacts the environment adversely as it can be toxic, ignitable, reactive and radioactive. Such waste pollutes the air, soil and nearby water sources thereby affecting the marine wildlife and the ecosystem. This chapter delves into the impacts of industrial waste to provide an in-depth understanding of the subject.

Chapter – Industrial Waste Management and Treatment

The process of managing waste released by various industries is termed as industrial waste management. Its management and treatment methods include landfill, composting, aerobic and anaerobic treatment system, zero-sludge treatment, etc. This chapter closely examines these methods and related concepts of industrial waste management and treatment to provide an extensive understanding of the subject.

Chapter - Industrial Wastewater Treatment

Different methods of treatment of industrial waste include advanced oxidation processing, distillation, adsorption, ozonation, vitrification, incineration, chemical immobilization and landfill disposal. This chapter has been carefully written to provide an easy understanding of these treatment processes of industrial waste.

Maria Malone

CHAPTER 1

The waste produced by the industries is referred to as industrial waste. It can be classified into toxic waste, hazardous waste, radioactive waste, household hazardous waste, chemical waste, solid waste, etc. This is an introductory chapter which will briefly introduce all the significant aspects of industrial wastes and its types.

INDUSTRIAL WASTE

Industrial waste is defined as waste generated by manufacturing or industrial processes. The types of industrial waste generated include cafeteria garbage, dirt and gravel, masonry and concrete, scrap metals, trash, oil, solvents, chemicals, weed grass and trees, wood and scrap lumber, and similar wastes. Industrial solid waste - which may be solid, liquid or gases held in containers - is divided into hazardous and non-hazardous waste. Hazardous waste may result from manufacturing or other industrial processes. Certain commercial products such as cleaning fluids, paints or pesticides discarded by commercial establishments or individuals can also be defined as hazardous waste. Non-hazardous industrial wastes are those that do not meet the EPA's definition of hazardous waste - and are not municipal waste. Industrial waste has been a problem since the industrial revolution. Industrial waste may be toxic, ignitable, corrosive or reactive. If improperly managed, this waste can pose dangerous health and environmental consequences. In the United States, the amount of hazardous waste generated by manufacturing industries in the country has increased from an estimated 4.5 million tons annually after World War II to some 57 million tons by 1975. By 1990, this total had shot up to approximately 265 million tons. This waste is generated at every stage in the production process, use and disposal of manufactured products.

Thus, the introduction of many new products for the home and office - computers, drugs, textiles, paints and dyes, plastics - also introduced hazardous waste, including toxic chemicals, into the environment. These, too, must be managed with extreme care to avoid adverse environmental or human health impacts. The EPA estimated in 1980 that more than 70,000 different chemicals were being manufactured in the U.S., with some 1,000 new chemicals being added each year. The human health and environmental impacts of many of these chemicals are largely unknown. High levels of toxic contaminants have been found in animals and humans, particularly those, like farm workers and oil and gas workers, who are continually exposed to such waste streams. Waste water from manufacturing or chemical processes in industries contributes to water pollution. Industrial waste water usually contains specific and readily identifiable chemical compounds. Water pollution is concentrated within a few subsectors, mainly in the form of toxic wastes and organic pollutants. Out of this a large portion can be traced to the processing of industrial chemicals and to the food products industry. Most major industries have treatment facilities for industrial effluents but this is not the case with small-scale industries, which cannot afford enormous investments in pollution control equipment as their profit margin is very slender. The effects of water pollution are not only devastating to people but also to animals, fish, and birds. Polluted water is unsuitable for drinking, recreation, agriculture, and industry. It diminishes the aesthetic quality of lakes and rivers. More seriously, contaminated water destroys aquatic life and reduces its reproductive ability. Eventually, it is a hazard to human health. Nobody can escape the effects of water pollution.

Two types of industrial waste of particular concern are dry cleaning fluids and embalming fluids. Dry cleaning fluids have contaminated groundwater supplies in all areas of the United States. One of the most common contaminants is PCE (perchloroethylene or tetrachloroethylene, $Cl_2C{=}CCl_2$). As a suspected carcinogen, PCE must be removed from water to very low levels. The US EPA maximum contaminant level (MCL) for PCE in drinking water is 5 ppb (5 parts per billion, or 5 mg/L). States such as New Jersey have set MCLs as low as 1 ppb for PCE in public water supplies. Cemeteries can be a source of ground water contamination caused by decaying organic matter and embalming fluids. In fact, there are several historical accounts of pollution of water wells in the vicinity of cemeteries. Embalming fluids include carcinogens. The threat of embalming fluids contaminating the water supply has prompted some cities to decrease the size of planned large cemeteries. With our aging population there is no doubt that embalming fluids will increasingly be a source of water pollution unless something changes.

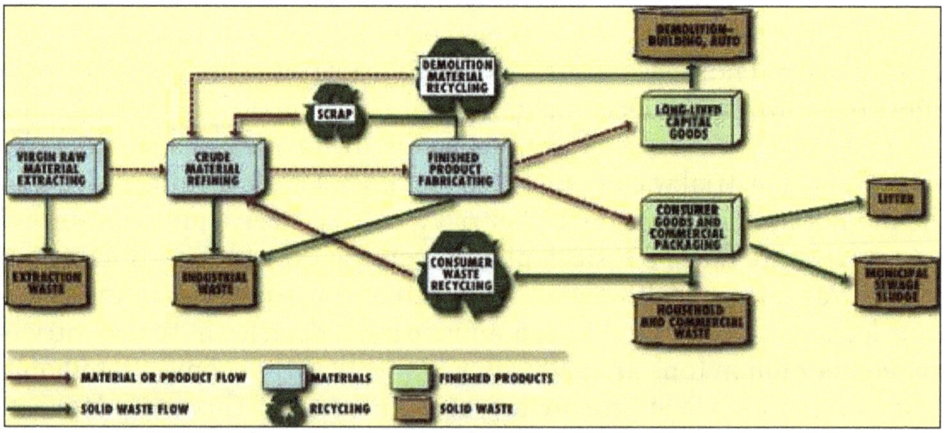

Flow of Materials, Products, and Solid Waste.

Examples of hazardous waste generated by industries and businesses	
Waste Generator	Waste Types
Chemical Manufactures	Acids And Bases. Spent Solvents. Reactive Waste. Wastewater Containing. Organic Constituents.
Printing Industry	Heavy Metal Solutions. Waste Inks. Solvents. Ink Sludge Containing Heavy Metals.

Petroleum Refining Industry	Wastewater Containing. Benzene and Other. Hydrocarbons. Sludge from Refining Process.
Leather Products Manufacturing	Toluene and Benzene
Paper Industry	Paint Waste Containing. Heavy Metals. Ignitable Solvents.
Construction Industry	Ignitable Paint Waste. Spent Solvents. Strong Acids and Bases.
Metal Manufacturing	Sludge Containing Heavy Metals. Cyanide Waste. Paint Waste.

What can be Done and what is being Done?

There are many ways to reduce waste in an industrial setting. Waste minimization includes any source reduction and recycling activity undertaken by a waste generator (i.e. any business that produces waste through their operations). These activities result in a reduction of waste produced and a reduction in the toxicity of the waste. Some examples of waste minimization techniques are listed below.

Waste reduction techniques:

- Change the composition of the product to reduce the amount of waste resulting from the product's use.
- Reduce or eliminate hazardous materials that enter the production process.
- Use technology (including measuring and cutting) to make changes to the production process; equipment, layout or piping; or operating conditions.
- Purchase what you need to avoid waste from unwanted materials.

Good operating practices such as waste minimization programs, management and personnel practices, loss prevention, and waste segregation help to reduce waste at their source.

Recycling Techniques:

- Return waste material to original process.
- Use the waste material as a raw material substitute for another process.
- Process waste material for resource recovery.
- Process waste material as a by-product.
- Investigate contractors to recycle waste material.

- Advertise waste material.
- Use packaging waste again (cardboard, bubble wrap or polystyrene).

The term Cleaner Production was coined by the United Nations Environment Program (UNEP) when it launched the Cleaner Production Program in 1989. Cleaner Production is the continuous application of an integrated preventive environmental strategy applied to processes, products, and services to increase overall efficiency and reduce risks to humans and the environment.

- For production processes: The strategy includes conserving raw materials and energy, eliminating toxic raw materials, and reducing the quantity and toxicity of all emissions and wastes.
- For products: The strategy focuses on reducing negative impacts along the life cycle of a product, from raw materials extraction to its ultimate disposal.
- For services: The strategy involves incorporating environmental concerns into designing and delivering services.
- Cleaner production requires changing attitudes, responsible environmental management and evaluating technology options.

Benefits of cleaner production:

- Economic benefits include increased profits; improved marketability; lower expenditure on materials and resources; additional sources of income; increased quality and greater productivity; better customer service; and increased goodwill from being seen as a responsible member of the community.
- Environmental benefits include less pollution; reduced production of waste; more efficient use of resources; less waste going to landfill; and less effluent to sewage or discharges to water bodies.
- The Ministry of Environment in each province regulates the discharges of solid, liquid and gaseous wastes to the environment from industrial sources. They do this through developing policies and regulations using the best available science, conducting stakeholder consultations, administering the authorizations associated with discharges to the environment, and conducting compliance activities.
- Technologies have been developed to solve a wide range of solid waste and other environmental and recycling problems. Many new technologies for separation and recycling of fly ash, metals, plastics, glass, sands, abrasives, automobile scrap, and plating waste sludge have been developed.

INDUSTRIAL SOLID WASTE

The major generators of industrial solid wastes are the thermal power plants producing coal ash, the integrated Iron and Steel mills producing blast furnace slag and steel melting slag, non-ferrous industries like aluminum, zinc and copper producing red mud and tailings, sugar industries generating press mud, pulp and paper industries producing lime and fertilizer and allied industries producing gypsum.

Table: Source and Quantum of generation of some major industrial waste.

S.No	Name	Quantity (million tonnes per annum)	Source/Origin
1.	Steel and Blast furnace	35.0	Conversion of pig iron to steel and manufacture of iron.
2.	Brine mud	0.02	Caustic soda industry.
3.	Copper slag	0.0164	By product from smelting of copper.
4.	Fly ash	70.0	Coal based thermal power plants.
5.	Kiln dust	1.6	Cement plants.
6.	Lime sludge	3.0	Sugar, paper, fertilizer tanneries, soda ash, calcium carbide industries.
7.	Mica scraper waste	0.005	Mica mining areas.
8.	Phosphogypsum	4.5	Phosphoric acid plant, Ammonium phosphate.
9.	Red mud/Bauxite	3.0	Mining and extraction of alumina from bauxite.
10.	Coal washery dust	3.0	Coal mines.
11.	Iron tailing	11.25	Iron ore.
12.	Lime stone wastes	50.0	Lime stone quarry.

Important Industrial Solid Waste

Coal Ash

In general, a 1,000 MW station using coal of 3,500 kilo calories per kg and ash content in the range of 40-50 per cent would need about 500 hectares for disposal of fly ash for about 30 years' operation. It is, therefore, necessary that fly ash should be utilised wherever possible to minimize environmental degradation. The thermal power plant should take into account the capital and operation/maintenance cost of fly ash disposal system as well as the associated environmental protection cost, vis-a-vis dry system of collection and its utilisation by the thermal power plant or other industry, in evaluating the feasibility of such system.

The research and development carried out for utilisation of fly ash for making building materials has proved that fly ash can be successfully utilised for production of bricks, cement and other building materials. Indigenous technology for construction of building materials utilising fly ash is available and are being practised in a few industries. However, large scale utilisation is yet to take off. Even if the full potential of fly ash utilisation through manufacture of fly ash bricks and blocks is explored, the quantity of fly ash produced by the thermal power plants are so huge that major portion of it will still remain unutilised. Hence, there is a need to evolve strategies and plans for safe and environmentally sound method of disposal.

Integrated Iron and Steel Plant Slag

The Blast Furnace (BF) and Steel Melting Shop (SMS) slags in integrated iron and steel plants are at present dumped in the surrounding areas of the steel plants making hillocks encroaching on the agricultural land. Although, the BF slag has potential for conversion into granulated slag, which is a useful raw material in cement manufacturing, it is yet to be practised in a big way. Even the use of slag as road subgrade or land-filling is also very limited.

Phosphogypsum

Phosphogypsum is the waste generated from the phosphoric acid, ammonium phospate and hydrofluoric acid plants. This is very useful as a building material. At present very little attention has been paid to its utilisation in making cement, gypsum board, partition panel, ceiling tiles, artificial marble, fiber boards etc.

Red Mud

Red mud as solid waste is generated in non-ferrous metal extraction industries like aluminum and copper. The red mud at present is disposed in tailing ponds for settling, which more often than not finds its course into the rivers, especially during monsoon. However, red mud has recently been successfully tried and a plant has been set up in the country for making corrugated sheets. Demand for such sheet should be popularised and encouraged for use. This may replace asbestos which is imported and also banned in developed countries for its hazardous effect. Attempts are also made to manufacture polymer and natural fibres composite panel doors from red mud.

Lime Mud

Lime sludge, also known as lime mud, is generated in pulp and paper mills which is not recovered for reclamation of calcium oxide for use except in the large mills. The lime mud disposal by dumping into low-lying areas or into water courses directly or as run-off during monsoon is not only creating serious pollution problem but also wasting the valuable non-renewable resources. The reasons for not reclaiming the calcium oxide in the sludge after recalcination is that it contains high amount of silica. Although a few technologies have been developed to desilicate black liquor before burning, none of the mills in the country are adopting desilication technology.

Waste Sludge and Residues

Treatment of industrial wastes/effluents results in generation of waste sludge/residues which, if not properly disposed, may cause ground and surface water pollution.

TOXIC WASTE

Toxic waste is a chemical waste material capable of causing death or injury to life. Waste is considered toxic if it is poisonous, radioactive, explosive, carcinogenic (causing cancer), mutagenic (causing damage to chromosomes), teratogenic (causing birth defects), or bioaccumulative (that is, increasing in concentration at the higher ends of food chains). Waste containing dangerous pathogens, such as used syringes, is sometimes considered to be toxic waste. Poisoning occurs when toxic waste is ingested, inhaled, or absorbed by the skin.

Toxic waste results from industrial, chemical, and biological processes. Toxins are found in household, office, and commercial wastes. Examples of common products that routinely become part of the toxic waste streams of industrialized countries include batteries for electronic devices, pesticides, cell phones, and computers. The U.S. Environmental Protection Agency estimated that U.S. factories released 1.8 million metric tons (about 2 million tons) of toxic chemicals into the air, land, and surface waters in 2011, including a number of chemicals that are known carcinogens. In the United States hundreds of billions of gallons of groundwater are also contaminated with uranium and other toxic chemicals, and more than 63.5 million metric tons (about 70 million tons) of radioactive waste, which is mostly uranium waste derived from spent nuclear fuel, is buried in landfills, trenches, and unlined tanks.

Several social and ethical issues permeate the discussion of toxic waste. In countries with lax pollution regulations where polluters have no incentive to limit the disposal of toxins in the air, water, or landfills, negative externalities (costs imposed on society at large but not borne by the polluter) exist; such a shifting of costs raises fundamental questions of fairness. In countries with more stringent pollution regulations, toxic wastes may be dumped illegally, and some polluters may attempt to cover up that activity. Another approach taken to dealing with toxic waste is to send it elsewhere; much electronic waste produced in the U.S. is shipped to developing countries, risking spillages and the health of local residents, who often lack the expertise and technology to safely deal with toxic waste. In addition, the practice of siting toxic waste storage or handling facilities in minority enclaves in some countries is considered by some environmentalists to be a form of environmental racism, the disproportionate shifting of environmental hazards to people of colour.

Types Toxic Waste

Toxic waste products are divided into three general categories: chemical wastes, radioactive wastes, and medical wastes. Chemical wastes, such as those that are considered corrosive, flammable, reactive (that is, chemicals that interact with others to create explosive or toxic by-products), acutely poisonous, carcinogenic, mutagenic, and tetratogenic—as well as heavy metals (such as lead and mercury)—are placed in the first category. Radioactive wastes include elements and compounds that produce or absorb ionizing radiation and any material that interacts with such elements and compounds (such as the rods and water that moderate nuclear reactions in power plants). Medical wastes are a broad category, spanning the range from tissues and fluids capable of harbouring infectious disease-causing organisms to the materials and containers that hold and transfer them.

The world's most dangerous chemical toxins, which are commonly grouped into a collection called the "dirty dozen" by chemists and environmentalists, are categorized as persistent organic pollutants (POPs). Several POPs are pesticides: Aldrin, chlordane, DDT, dieldrin, endrin, heptachlor, hexachlorobenzene, mirex, and toxaphene. Other POPs are produced during the combustion process. For example, dioxins and furans are by-products of chemical production and the burning of chlorinated substances, and polychlorinated biphenyls (PCBs), which are used to manufacture such products as paints, plastics, and electrical transformers, may be released into the air when those products are burned. Other toxins such as arsenic, beryllium, cadmium, copper, lead, nickel, and zinc belong to a wider group of chemicals called persistent bioaccumulative toxins (PBTs), which include the dirty dozen and can linger in the environment for long periods.

Hazards

Well before the 1962 publication of American biologist Rachel Carson's Silent Spring, which described how DDT accumulated in the fatty tissues of animals and caused cancer and genetic damage, the risks of many toxic wastes were evident. For example, lead was a known toxin in the 19th century, with reformers documenting lead poisoning in the workforce and leading cleanup efforts. Nevertheless, auto companies, oil companies, and the U.S. government authorized the manufacture, distribution, and use of tetraethyl lead, $Pb(C_2H_5)_4$, in gasoline in the 1920s. Health officials warned against depositing millions of pounds of inorganic lead dust from automobile exhaust onto the streets. However, the lead industry pointed to lead's importance to the automotive and petrochemical industries in increasing engine performance and reducing engine knock (spontaneous ignition of the fuel-air mixture in vehicle engines). Similarly, despite evidence of lead paint's toxic effects on children as early as the 1920s, the lead industry campaigned for decades to deter concerns. The National Lead Company, manufacturer of Dutch Boy paints and lead pigments, produced children's colouring books, including The Dutch Boy's Lead Party, extolling the benefits of lead paint. The federal government finally banned lead in paint and gasoline in the 1970s and '80s.

Although limited cases of accidental poisonings, such as from the accidental ingestion of lead and household cleaners, occur daily throughout the world, one of the first high-profile episodes of mass poisonings affecting neighbourhoods and whole cities occurred in Minamata, Japan, in the 1950s. Many of the town's residents contracted mercury poisoning resulting from the Nippon Chisso Hiryo Co.'s manufacturing of acetaldehyde, and that material was later associated with the deaths of at least 3,000 people. Mercury from the production process spilled into the bay and entered the food chain, including seafood, which was the town's primary protein source. Deformed fish appeared in Minamata Bay, and townspeople exhibited strange behaviours, including trembling, stumbling, uncontrollable shouting, paralysis, hearing and vision problems, and body contortions. While mercury was long known to be a toxin (the neurological degeneration caused by mercury used in hat making in the 19th century led to the phrase "mad as a hatter"), Minamata vividly highlighted its dangers in the food chain.

Hooker Chemical and Plastics Corporation used an empty canal in Love Canal, a section of Niagara Falls, New York, in the 1940s and '50s to dump 20,000 tons of toxic waste in metal drums. After the canal was filled and the land given to the city, houses and an elementary school were built on the site. By the late 1970s the toxic chemicals had leaked through their drums and risen to the surface, resulting in high rates of birth defects, miscarriages, cancer and other illnesses, and chromosome damage. The neighbourhood was subsequently evacuated by September 1979.

Dust from the remains of the three World Trade Center buildings that were destroyed during the September 11, 2001, terrorist attacks in New York were found to contain mercury, lead, dioxin, and asbestos. Aside from the dangers of breathing in toxic building materials, the attacks raised concerns about potential sabotage of toxic waste sites, such as storage facilities adjacent to nuclear power plants, or of the transport of such waste between sites. More than 15,000 chemical plants and refineries nationwide were also in danger, with more than 100 of them putting at least a million people at risk should an attack occur.

In addition, the danger of a sudden release of toxic material also looms in the aftermath of extreme weather events, natural disasters, and accidents. Three Superfund toxic waste sites in and around New Orleans were flooded in 2005 by Hurricane Katrina, and toxic waste was found in debris deposited throughout the flooded area. The devastating Indian Ocean earthquake and tsunami of 2004 stirred up and dispersed vast amounts of toxic wastes—including radioactive waste, lead, heavy metals, and hospital wastes—across the Indian Ocean basin, and the tsunami that struck Japan in 2011, which caused the Fukushima nuclear accident, released tremendous amounts of irradiated water into the Pacific Ocean. Those and other high-profile examples—including the Exxon Valdez oil spill in 1989, the Chernobyl disaster in 1986, the Bhopal gas leak in 1985, and the Three-Mile Island scare in 1979—raised public awareness and concern.

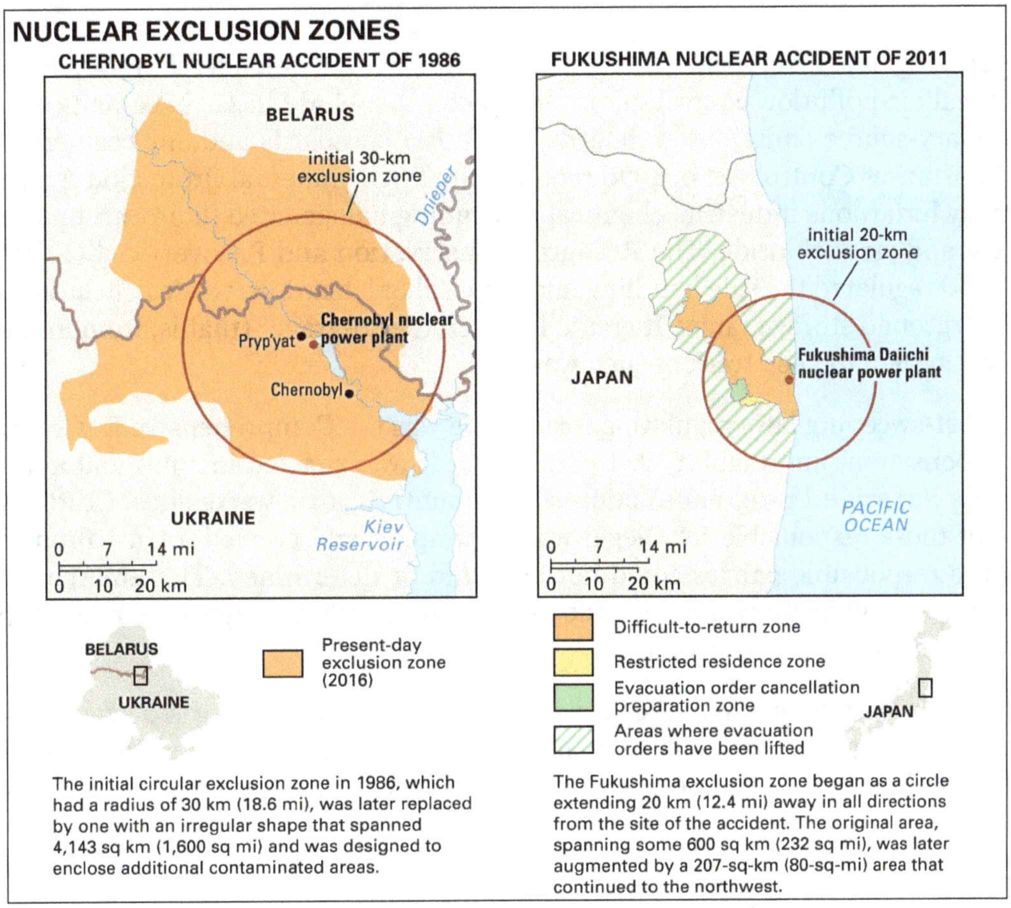

Exclusion zone: Chernobyl disaster; Fukushima accident.

Map of the exclusion zones following the nuclear accidents at Chernobyl, Soviet Union (now in Ukraine), and Fukushima, Japan.

Costs

Toxic wastes result in huge costs in terms of economic expenditures, human health, and ecosystem health. The U.S. Geological Survey places cleanup costs for existing environmental contamination in the United States between several hundred million and more than one trillion dollars. Toxic waste has been implicated in deaths and health problems such as cancers, birth defects, miscarriages, low birth weight, neurological disorders, liver disease, developmental disorders, hypertension, and heart defects.

In ecosystems, toxic wastes have caused substantial damage to animal and plant populations. Such wastes overwhelm natural restorative processes, destroy habitats, and reduce populations of sensitive species outright or inhibit their reproductive success. The decline of the bald eagle population by the 1960s and the double-crested cormorant population by the early 1970s as a result of DDT use are vivid examples. Similarly, PCBs and other toxins are blamed for many whale deaths, bird deformities, and endocrine, reproductive, neurological, and immune system disruptions in humans and wildlife.

Laws

Many U.S. laws regulate toxic waste. The 1970 Clean Air Act, last amended in 1990, forms the basis for the national air-pollution control effort. Its elements include hazardous air-pollutants standards, stationary-source emissions standards, and other standards and enforcement provisions. The Toxic Substances Control Act of 1976 requires the Environmental Protection Agency to regulate potentially hazardous industrial chemicals, including halogenated fluorocarbons, dioxins, asbestos, PCBs, and vinyl chloride. The Resource Conservation and Recovery Act (RCRA) became law in 1976 and regulated the safe handling and disposal of hazardous wastes, including those that occur in underground storage tanks. It created the "cradle-to-grave" (that is, from manufacture to final disposal) system to keep track of such wastes.

One of the most-sweeping laws regulating toxic waste was the Comprehensive Environmental Response, Compensation, and Liability Act (CERCLA). It was enacted in 1980 and authorized the creation of the Superfund program to address the country's toxic waste sites. CERCLA provided for liability of those responsible for illegal waste dumping and created a trust fund to clean up sites when the responsible parties could not be found or determined. The Emergency Planning and Community Right-to-Know Act amended CERCLA in 1986 to require mandatory public disclosure of any release of toxic substances.

In addition to such regulations were the growing numbers of "toxic tort" cases against producers of toxic waste. A toxic tort is personal injury or property damage from exposure to toxic substances due to the fault of another party. Victims can sue for medical expenses, lost wages, and pain and suffering.

HAZARDOUS WASTE

Hazardous waste is waste that has substantial or potential threats to public health or the environment.

- Characteristic hazardous wastes are materials that are known or tested to exhibit one or more of the following hazardous traits:
    - Ignitability.

  - Reactivity.
  - Corrosivity.
  - Toxicity.

- Listed hazardous wastes are materials specifically listed by regulatory authorities as hazardous wastes which are from non-specific sources, specific sources, or discarded chemical products.

Hazardous wastes may be found in different physical states such as gaseous, liquids, or solids. A hazardous waste is a special type of waste because it cannot be disposed of by common means like other by-products of our everyday lives. Depending on the physical state of the waste, treatment and solidification processes might be required.

Worldwide, the United Nations Environment Programme (UNEP) estimated that more than 400 million tons of hazardous wastes are produced universally each year, mostly by industrialized countries. About 1 percent of this is shipped across international boundaries, with the majority of the transfers occurring between countries in the Organization for the Economic Cooperation and Development(OECD). One of the reasons for industrialized countries to ship the hazardous waste to industrializing countries for disposal is the rising cost of disposing of hazardous waste in the home country.

In the United States, the treatment, storage, and disposal of hazardous waste are regulated under the Resource Conservation and Recovery Act (RCRA). Hazardous wastes are defined under RCRA in 40 CFR 261 where they are divided into two major categories: Characteristic wastes and listed wastes.

The requirements of the RCRA apply to all the companies that generate hazardous waste as well as those companies that store or dispose hazardous waste in the United States. Many types of businesses generate hazardous waste. Dry cleaners, automobile repair shops, hospitals, exterminators, and photo processing centers may all generate hazardous waste. Some hazardous waste generators are larger companies such as chemical manufacturers, electroplating companies, and oil refineries.

A U.S. facility that treats, stores, or disposes of hazardous waste must obtain a permit for doing so under the Resource Conservation and Recovery Act. Generators and transporters of hazardous waste must meet specific requirements for handling, managing, and tracking waste. Through the RCRA, Congress directed the United States Environmental Protection Agency (EPA) to create regulations to manage hazardous waste. Under this mandate, the EPA developed strict requirements for all aspects of hazardous waste management including the treatment, storage, and disposal of hazardous waste. In addition to these federal requirements, states may develop more stringent requirements that are broader in scope than the federal regulations. Furthermore, RCRA allows states to develop regulatory programs that are at least as stringent as RCRA and, after review by EPA, the states may take over responsibility for the implementation of the requirements under RCRA. Most states take advantage of this authority, implementing their own hazardous waste programs that are at least as stringent, and in some cases are more stringent than the federal program.

Hazardous Waste Mapping Systems

The U.S. government provides several tools for mapping hazardous wastes to particular locations. These tools also allow the user to view additional information.

- TOXMAP is a Geographic Information System (GIS) service from the Division of Specialized Information Services of the United States National Library of Medicine (NLM) that uses maps of the United States to help users visually explore data from the United States Environmental Protection Agency's (EPA) Toxics Release Inventory and Superfund Basic Research Program. This is a resource funded by the US Federal Government. TOXMAP's chemical and environmental health information is taken from NLM's Toxicology Data Network (TOXNET), PubMed, and other authoritative sources.
- The US Environmental Protection Agency (EPA) "Where You Live" allows users to select a region from a map to find information about Superfund sites in that region.

Universal Wastes

Universal wastes are a special category of hazardous wastes that:

- Generally pose a lower threat relative to other hazardous wastes are ubiquitous and produced in very large quantities by a large number of generators.

Some of the most common "universal wastes" are: Fluorescent light bulbs, some specialty batteries (e.g. lithium or lead containing batteries), cathode ray tubes, and mercury-containing devices.

Universal wastes are subject to somewhat less stringent regulatory requirements. Small quantity generators of universal wastes may be classified as "conditionally exempt small quantity generators" (CESQGs) which release them from some of the regulatory requirements for the handling and storage hazardous wastes.

Universal wastes must still be disposed of properly.

Disposal of Hazardous Waste

Historically, some hazardous wastes were disposed of in regular landfills. This resulted in unfavorable amounts of hazardous materials seeping into the ground. These chemicals eventually entered to natural hydrologic systems. Many landfills now require countermeasures against groundwater contamination. For example, a barrier has to be installed along the foundation of the landfill to contain the hazardous substances that may remain in the disposed waste. Currently, hazardous wastes must often be stabilized and solidified in order to enter a landfill and must undergo different treatments in order to stabilize and dispose of them. Most flammable materials can be recycled into industrial fuel. Some materials with hazardous constituents can be recycled, such as lead acid batteries.

Recycling

Some hazardous wastes can be recycled into new products. Examples may include lead-acid batteries or electronic circuit boards. When heavy metals in these types of ashes go through the proper treatment, they could bind to other pollutants and convert them into easier-to-dispose solids, or they could be used as pavement filling. Such treatments reduce the level of threat of harmful chemicals, like fly and bottom ash, while also recycling the safe product. There is a recycling center facility in Oxnard, CA. The city does not charge for any hazardous materials being disposed of, but there is a limit to how much you can bring per month. Other than hazardous waste, the city also allows you to dispose of electronic waste, light-bulbs, and batteries.

Portland Cement

Another commonly used treatment is cement-based solidification and stabilization. Cement is used because it can treat a range of hazardous wastes by improving physical characteristics and decreasing the toxicity and transmission of contaminants. The cement produced is categorized into 5 different divisions, depending on its strength and components. This process of converting sludge into cement might include the addition of pH adjustment agents, phosphates, or sulfur reagents to reduce the settling or curing time, increase the compressive strength, or reduce the leach ability of contaminants.

Incineration, Destruction and Waste-to-energy

Hazardous waste may be "destroyed". For example, by incinerating them at a high temperature, flammable wastes can sometimes be burned as energy sources. For example, many cement kilns burn hazardous wastes like used oils or solvents. Today, incineration treatments not only reduce the amount of hazardous waste, but also generate energy from the gases released in the process. It is known that this particular waste treatment releases toxic gases produced by the combustion of byproduct or other materials which can affect the environment. However, current technology has developed more efficient incinerator units that control these emissions to a point where this treatment is considered a more beneficial option. There are different types of incinerators which vary depending on the characteristics of the waste. Starved air incineration is another method used to treat hazardous wastes. Just like in common incineration, burning occurs, however controlling the amount of oxygen allowed proves to be significant to reduce the amount of harmful byproducts produced. Starved air incineration is an improvement of the traditional incinerators in terms of air pollution. Using this technology, it is possible to control the combustion rate of the waste and therefore reduce the air pollutants produced in the process.

Hazardous Waste Landfill

Hazardous waste may be sequestered in an hazardous waste landfill or permanent disposal facility. "In terms of hazardous waste, a landfill is defined as a disposal facility or part of a facility where hazardous waste is placed or on land and which is not a pile, a land treatment facility, a surface impoundment, an underground injection well, a salt dome formation, a salt bed formation, an underground mine, a cave, or a corrective action management unit (40 CFR 260.10)."

Pyrolysis

Some hazardous waste types may be eliminated using pyrolysis in a high temperature not necessarily through electrical arc but starved of oxygen to avoid combustion. However, when electrical arc is used to generate the required ultra heat (in excess of 3000 degree C temperature) all materials (waste) introduced into the process will melt into a molten slag and this technology is termed Plasma not pyrolysis. Plasma technology produces inert materials and when cooled solidifies into rock like material. These treatment methods are very expensive but may be preferable to high temperature incineration in some circumstances such as in the destruction of concentrated organic waste types, including PCBs, pesticides and other persistent organic pollutants.

RADIOACTIVE WASTE

Tint low-level radioactive waste barrels.

Radioactive waste is a type of hazardous waste that contains radioactive material. Radioactive waste is usually a by-product of nuclear power generation and other applications of nuclear fission or nuclear technology, such as research and medicine. Radioactive waste is regulated by government agencies in order to protect human health and the environment.

Radioactivity naturally decreases over time, so radioactive waste has to be isolated and confined in appropriate disposal facilities for a sufficient period until it no longer poses a threat. The time radioactive waste must be stored for depends on the type of waste and radioactive isotopes. Current approaches to managing radioactive waste have been segregation and storage for short-lived waste, near-surface disposal for low and some intermediate-level waste, and burial in a deep geological repository or transmutation for the high-level waste.

Nature and Significance

Radioactive waste typically comprises a number of radionuclides: Unstable configurations of elements that decay, emitting ionizing radiation which is harmful to humans and the environment. These isotopes emit different types and levels of radiation, which last for different periods of time.

Physics

Medium-lived fission products				
Prop: Unit:	t½ (a)	Yield (%)	Q * (keV)	βγ *
^{155}Eu	4.76	0.0803	252	βγ
^{85}Kr	10.76	0.2180	687	βγ
^{113m}Cd	14.1	0.0008	316	β
^{90}Sr	28.9	4.505	2826	β
^{137}Cs	30.23	6.337	1176	βγ
^{121m}Sn	43.9	0.00005	390	βγ
^{151}Sm	88.8	0.5314	77	β

Long-Lived Fission Products				
Nuclide	t½	Yield	Decay energy[a 1]	Decay mode
	(Ma)	(%)[a 2]	(keV)	
^{99}Tc	0.211	6.1385	294	β
^{126}Sn	0.230	0.1084	4050[a 3]	βγ
^{79}Se	0.327	0.0447	151	β
^{93}Zr	1.53	5.4575	91	βγ
^{135}Cs	2.3	6.9110[a 4]	269	β
^{107}Pd	6.5	1.2499	33	β
^{129}I	15.7	0.8410	194	βγ

1. ^ Decay energy is split among β, neutrino, and γ if any.
2. ^ Per 65 thermal-neutron fissions of U-235 and 35 of Pu-239.
3. ^Has decay energy 380 keV, but decay product Sb-126 has decay energy 3.67 MeV.
4. ^ Lower in thermal reactor because predecessor absorbs neutrons.

The radioactivity of all radioactive waste weakens with time. All radionuclides contained in the waste have a half-life — the time it takes for half of the atoms to decay into another nuclide — and eventually, all radioactive waste decays into non-radioactive elements (i.e., stable nuclides). Since radioactive decay follows the half-life rule, the rate of decay is inversely proportional to the duration of decay. In other words, the radiation from a long-lived isotope like iodine-129 will be much less intense than that of a short-lived isotope like iodine-131. The two tables show some of the major radioisotopes, their half-lives, and their radiation yield as a proportion of the yield of fission of uranium-235.

The energy and the type of the ionizing radiation emitted by a radioactive substance are also important factors in determining its threat to humans. The chemical properties of the radioactive element will determine how mobile the substance is and how likely it is to spread into the environment and contaminate humans. This is further complicated by the fact that many radioisotopes do not decay immediately to a stable state but rather to radioactive decay products within a decay chain before ultimately reaching a stable state.

Pharmacokinetics

Exposure to radioactive waste may cause health impacts due to ionizing radiation exposure. In humans, a dose of 1 sievert carries a 5.5% risk of developing cancer, and regulatory agencies assume the risk is linearly proportional to dose even for low doses. Ionizing radiation can cause deletions in chromosomes. If a developing organism such as a fetus is irradiated, it is possible a birth defect may be induced, but it is unlikely this defect will be in a gamete or a gamete-forming cell. The incidence of radiation-induced mutations in humans is small, as in most mammals, because of natural cellular-repair mechanisms, many just now coming to light. These mechanisms range from DNA, mRNA and protein repair, to internal lysosomic digestion of defective proteins, and even induced cell suicide—apoptosis.

Depending on the decay mode and the pharmacokinetics of an element (how the body processes it and how quickly), the threat due to exposure to a given activity of a radioisotope will differ. For instance iodine-131 is a short-lived beta and gamma emitter, but because it concentrates in the thyroid gland, it is more able to cause injury than caesium-137 which, being water soluble, is rapidly excreted through urine. In a similar way, the alpha emitting actinides and radium are considered very harmful as they tend to have long biological half-lives and their radiation has a high relative biological effectiveness, making it far more damaging to tissues per amount of energy deposited. Because of such differences, the rules determining biological injury differ widely according to the radioisotope, time of exposure and sometimes also the nature of the chemical compound which contains the radioisotope.

Sources

Radioactive waste comes from a number of sources. In countries with nuclear power plants, nuclear armament, or nuclear fuel treatment plants, the majority of waste originates from the nuclear fuel cycle and nuclear weapons reprocessing. Other sources include medical and industrial wastes, as well as naturally occurring radioactive materials (NORM) that can be concentrated as a result of the processing or consumption of coal, oil and gas, and some minerals.

Nuclear Fuel Cycle

Front End

Waste from the front end of the nuclear fuel cycle is usually alpha-emitting waste from the extraction of uranium. It often contains radium and its decay products.

Uranium dioxide (UO_2) concentrate from mining is a thousand or so times as radioactive as the granite used in buildings. It is refined from yellowcake (U_3O_8), then converted to uranium hexafluoride gas (UF_6). As a gas, it undergoes enrichment to increase the U-235 content from 0.7% to about 4.4% (LEU). It is then turned into a hard ceramic oxide (UO_2) for assembly as reactor fuel elements.

The main by-product of enrichment is depleted uranium (DU), principally the U-238 isotope, with a U-235 content of ~0.3%. It is stored, either as UF_6 or as U_3O_8. Some is used in applications where its extremely high density makes it valuable such as anti-tank shells, and on at least one occasion even a sailboat keel. It is also used with plutonium for making mixed oxide fuel (MOX) and to dilute, or downblend, highly enriched uranium from weapons stockpiles which is now being redirected to become reactor fuel.

Back End

The back-end of the nuclear fuel cycle, mostly spent fuel rods, contains fission products that emit beta and gamma radiation, and actinides that emit alpha particles, such as uranium-234 (half-life 245 thousand years), neptunium-237 (2.144 million years), plutonium-238 (87.7 years) and americium-241 (432 years), and even sometimes some neutron emitters such as californium (half-life of 898 years for Cf-251). These isotopes are formed in nuclear reactors.

It is important to distinguish the processing of uranium to make fuel from the reprocessing of used fuel. Used fuel contains the highly radioactive products of fission. Many of these are neutron absorbers, called neutron poisons in this context. These eventually build up to a level where they absorb so many neutrons that the chain reaction stops, even with the control rods completely removed. At that point the fuel has to be replaced in the reactor with fresh fuel, even though there is still a substantial quantity of uranium-235 and plutonium present. In the United States, this used fuel is usually "stored", while in other countries such as Russia, the United Kingdom, France, Japan and India, the fuel is reprocessed to remove the fission products, and the fuel can then be re-used. The fission products removed from the fuel are a concentrated form of high-level waste as are the chemicals used in the process.

Fuel Composition and Long Term Radioactivity

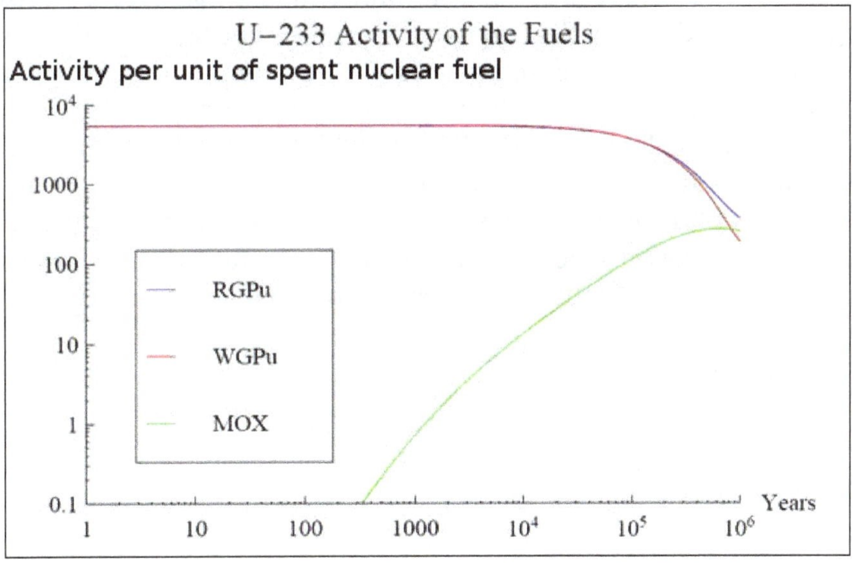

Activity of U-233 for three fuel types. In the case of MOX, the U-233 increases for the first 650 thousand years as it is produced by decay of Np-237 which was created in the reactor by absorption of neutrons by U-235.

The use of different fuels in nuclear reactors results in different spent nuclear fuel (SNF) composition, with varying activity curves.

Long-lived radioactive waste from the back end of the fuel cycle is especially relevant when designing a complete waste management plan for SNF. When looking at long-term radioactive decay, the actinides in the SNF have a significant influence due to their characteristically long half-lives. Depending on what a nuclear reactor is fueled with, the actinide composition in the SNF will be different.

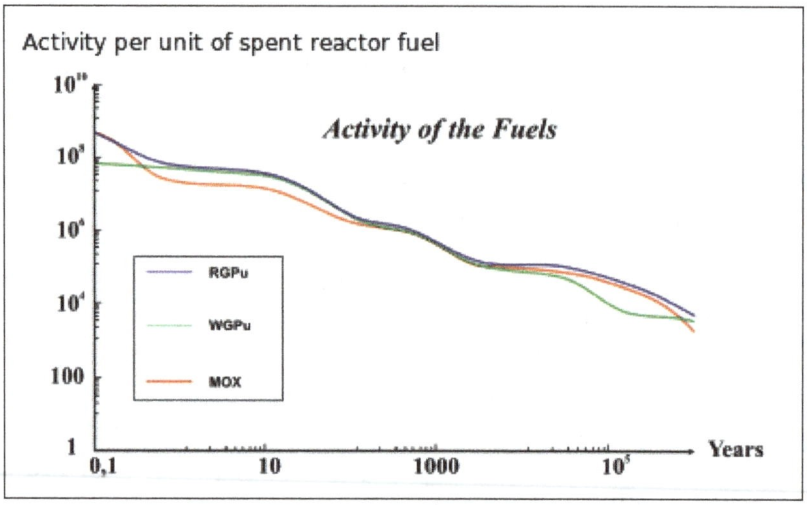

Total activity for three fuel types. In region 1 we have radiation from short-lived nuclides, and in region 2 from Sr-90 and Cs-137. On the far right we see the decay of Np-237 and U-233.

An example of this effect is the use of nuclear fuels with thorium. Th-232 is a fertile material that can undergo a neutron capture reaction and two beta minus decays, resulting in the production of fissile U-233. The SNF of a cycle with thorium will contain U-233. Its radioactive decay will strongly influence the long-term activity curve of the SNF around a million years. A comparison of the activity associated to U-233 for three different SNF types can be seen in the figure on the top right. The burnt fuels are thorium with reactor-grade plutonium (RGPu), thorium with weapons-grade plutonium (WGPu) and Mixed Oxide fuel (MOX, no thorium). For RGPu and WGPu, the initial amount of U-233 and its decay around a million years can be seen. This has an effect in the total activity curve of the three fuel types. The initial absence of U-233 and its daughter products in the MOX fuel results in a lower activity in region 3 of the figure on the bottom right, whereas for RGPu and WGPu the curve is maintained higher due to the presence of U-233 that has not fully decayed. Nuclear reprocessing can remove the actinides from the spent fuel so they can be used or destroyed.

Proliferation Concerns

Since uranium and plutonium are nuclear weapons materials, there have been proliferation concerns. Ordinarily (in spent nuclear fuel), plutonium is reactor-grade plutonium. In addition to plutonium-239, which is highly suitable for building nuclear weapons, it contains large amounts of undesirable contaminants: plutonium-240, plutonium-241, and plutonium-238. These isotopes are extremely difficult to separate, and more cost-effective ways of obtaining fissile material exist (e.g., uranium enrichment or dedicated plutonium production reactors).

High-level waste is full of highly radioactive fission products, most of which are relatively short-lived. This is a concern since if the waste is stored, perhaps in deep geological storage, over many years the fission products decay, decreasing the radioactivity of the waste and making the plutonium easier to access. The undesirable contaminant Pu-240 decays faster than the Pu-239, and thus the quality of the bomb material increases with time (although its quantity decreases during that time as well). Thus, some have argued, as time passes, these deep storage areas have the potential to become "plutonium mines", from which material for nuclear weapons can be acquired with relatively little difficulty. Critics of the latter idea have pointed out the difficulty of recovering useful material from sealed deep storage areas makes other methods preferable. Specifically, the high radioactivity and heat (80°C in surrounding rock) greatly increases the difficulty of mining a storage area, and the enrichment methods required have high capital costs.

Pu-239 decays to U-235 which is suitable for weapons and which has a very long half-life (roughly 10^9 years). Thus plutonium may decay and leave uranium-235. However, modern reactors are only moderately enriched with U-235 relative to U-238, so the U-238 continues to serve as a denaturation agent for any U-235 produced by plutonium decay.

One solution to this problem is to recycle the plutonium and use it as a fuel e.g. in fast reactors. In pyrometallurgical fast reactors, the separated plutonium and uranium are contaminated by actinides and cannot be used for nuclear weapons.

Nuclear Weapons Decommissioning

Waste from nuclear weapons decommissioning is unlikely to contain much beta or gamma activity other than tritium and americium. It is more likely to contain alpha-emitting actinides such as Pu-239 which is a fissile material used in bombs, plus some material with much higher specific activities, such as Pu-238 or Po.

In the past the neutron trigger for an atomic bomb tended to be beryllium and a high activity alpha emitter such as polonium; an alternative to polonium is Pu-238. For reasons of national security, details of the design of modern bombs are normally not released to the open literature.

Some designs might contain a radioisotope thermoelectric generator using Pu-238 to provide a long lasting source of electrical power for the electronics in the device.

It is likely that the fissile material of an old bomb which is due for refitting will contain decay products of the plutonium isotopes used in it, these are likely to include U-236 from Pu-240 impurities, plus some U-235 from decay of the Pu-239; due to the relatively long half-life of these Pu isotopes, these wastes from radioactive decay of bomb core material would be very small, and in any case, far less dangerous (even in terms of simple radioactivity) than the Pu-239 itself.

The beta decay of Pu-241 forms Am-241; the in-growth of americium is likely to be a greater problem than the decay of Pu-239 and Pu-240 as the americium is a gamma emitter (increasing external-exposure to workers) and is an alpha emitter which can cause the generation of heat. The plutonium could be separated from the americium by several different processes; these would include pyrochemical processes and aqueous/organic solvent extraction. A truncated PUREX type extraction process would be one possible method of making the separation. Naturally occurring uranium is not fissile because it contains 99.3% of U-238 and only 0.7% of U-235.

Legacy Waste

Due to historic activities typically related to radium industry, uranium mining, and military programs, numerous sites contain or are contaminated with radioactivity. In the United States alone, the Department of Energy states there are "millions of gallons of radioactive waste" as well as "thousands of tons of spent nuclear fuel and material" and also "huge quantities of contaminated soil and water." Despite copious quantities of waste, the DOE has stated a goal of cleaning all presently contaminated sites successfully by 2025. The Fernald, Ohio site for example had "31 million pounds of uranium product", "2.5 billion pounds of waste", "2.75 million cubic yards of contaminated soil and debris", and a "223 acre portion of the underlying Great Miami Aquifer had uranium levels above drinking standards." The United States has at least 108 sites designated as areas that are contaminated and unusable, sometimes many thousands of acres. DOE wishes to clean or mitigate many or all by 2025, using the recently developed method of geomelting, however the task can be difficult and it acknowledges that some may never be completely remediated. In just one of these 108 larger designations, Oak Ridge National Laboratory, there were for example at least "167 known contaminant release sites" in one of the three subdivisions of the 37,000-acre (150 km²) site. Some of the U.S. sites were smaller in nature, however, cleanup issues were simpler to address, and DOE has successfully completed cleanup, or at least closure, of several sites.

Medicine

Radioactive medical waste tends to contain beta particle and gamma ray emitters. It can be divided into two main classes. In diagnostic nuclear medicine a number of short-lived gamma emitters such as technetium-99m are used. Many of these can be disposed of by leaving it to decay for a short time before disposal as normal waste. Other isotopes used in medicine, with half-lives in parentheses, include:

- Y-90, used for treating lymphoma (2.7 days).
- I-131, used for thyroid function tests and for treating thyroid cancer (8.0 days).
- Sr-89, used for treating bone cancer, intravenous injection (52 days).
- Ir-192, used for brachytherapy (74 days).
- Co-60, used for brachytherapy and external radiotherapy (5.3 years).
- Cs-137, used for brachytherapy, external radiotherapy (30 years).

Industry

Industrial source waste can contain alpha, beta, neutron or gamma emitters. Gamma emitters are used in radiography while neutron emitting sources are used in a range of applications, such as oil well logging.

Naturally occurring Radioactive Material

Substances containing natural radioactivity are known as NORM (Naturally occurring radioactive material). After human processing that exposes or concentrates this natural radioactivity (such as mining bringing coal to the surface or burning it to produce concentrated ash), it becomes technologically enhanced naturally occurring radioactive material (TENORM). A lot of this waste is alpha particle-emitting matter from the decay chains of uranium and thorium. The main source of radiation in the human body is potassium-40 (^{40}K), typically 17 milligrams in the body at a time and

0.4 milligrams/day intake. Most rocks, due to their components, have a low level of radioactivity. Usually ranging from 1 millisievert (mSv) to 13 mSv annually depending on location, average radiation exposure from natural radioisotopes is 2.0 mSv per person a year worldwide. This makes up the majority of typical total dosage (with mean annual exposure from other sources amounting to 0.6 mSv from medical tests averaged over the whole populace, 0.4 mSv from cosmic rays, 0.005 mSv from the legacy of past atmospheric nuclear testing, 0.005 mSv occupational exposure, 0.002 mSv from the Chernobyl disaster, and 0.0002 mSv from the nuclear fuel cycle).

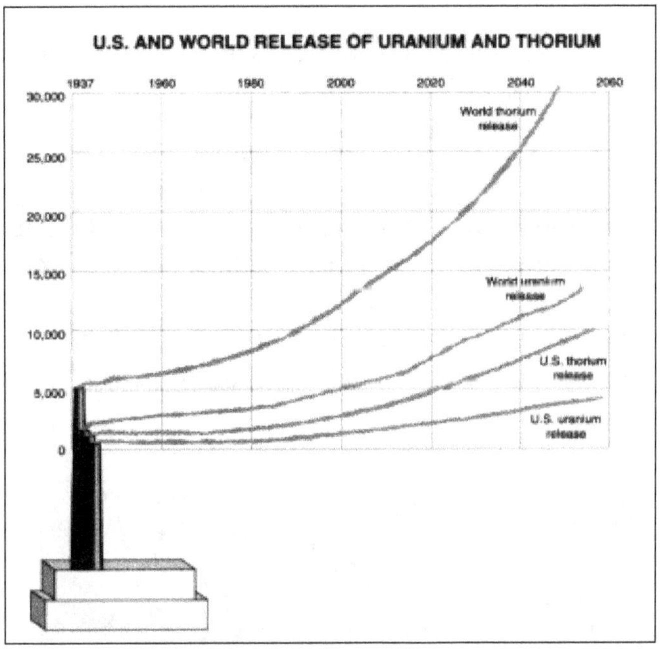

Annual release of uranium and thorium radioisotopes from coal combustion, predicted by ORNL to cumulatively amount to 2.9 million tons from the combustion of an estimated 637 billion tons of coal worldwide.

TENORM is not regulated as restrictively as nuclear reactor waste, though there are no significant differences in the radiological risks of these materials.

Coal

Coal contains a small amount of radioactive uranium, barium, thorium and potassium but in the case of pure coal, this is significantly less than the average concentration of those elements in the Earth's crust. The surrounding strata, if shale or mudstone, often contain slightly more than average and this may also be reflected in the ash content of 'dirty' coals. The more active ash minerals become concentrated in the fly ash precisely because they do not burn well. The radioactivity of fly ash is about the same as black shale and is less than phosphate rocks, but is more of a concern because a small amount of the fly ash ends up in the atmosphere where it can be inhaled. According to U.S. NCRP reports, population exposure from 1000-MWe power plants amounts to 490 person-rem/year for coal power plants, 100 times as great as nuclear power plants (4.8 person-rem/year). The exposure from the complete nuclear fuel cycle from mining to waste disposal is 136 person-rem/year; the corresponding value for coal use from mining to waste disposal is "probably unknown".

Oil and Gas

Residues from the oil and gas industry often contain radium and its decay products. The sulfate scale from an oil well can be very radium rich, while the water, oil and gas from a well often contain radon. The radon decays to form solid radioisotopes which form coatings on the inside of pipework. In an oil processing plant the area of the plant where propane is processed is often one of the more contaminated areas of the plant as radon has a similar boiling point to propane.

Classification

Classifications of radioactive waste varies by country. The IAEA, which publishes the Radioactive Waste Safety Standards (RADWASS), also plays a significant role.

Mill Tailings

Uranium tailings are waste by-product materials left over from the rough processing of uranium-bearing ore. They are not significantly radioactive. Mill tailings are sometimes referred to as 11(e)2 wastes, from the section of the Atomic Energy Act of 1946 that defines them. Uranium mill tailings typically also contain chemically hazardous heavy metal such as lead and arsenic. Vast mounds of uranium mill tailings are left at many old mining sites, especially in Colorado, New Mexico, and Utah.

Removal of very low-level waste.

Although mill tailings are not very radioactive, they have long half-lives. Mill tailings often contain radium, thorium and trace amounts of uranium.

Low-level Waste

Low level waste (LLW) is generated from hospitals and industry, as well as the nuclear fuel cycle. Low-level wastes include paper, rags, tools, clothing, filters, and other materials which contain small amounts of mostly short-lived radioactivity. Materials that originate from any region of an Active Area are commonly designated as LLW as a precautionary measure even if there is only a remote possibility of being contaminated with radioactive materials. Such LLW typically exhibits no higher radioactivity than one would expect from the same material disposed of in a non-active area, such as a normal office block. Example LLW includes wiping rags, mops, medical tubes, laboratory animal carcasses, and more.

Some high-activity LLW requires shielding during handling and transport but most LLW is suitable for shallow land burial. To reduce its volume, it is often compacted or incinerated before disposal. Low-level waste is divided into four classes: Class A, class B, class C, and Greater Than Class C (GTCC).

Intermediate-level Waste

Spent fuel flasks are transported by railway. Each flask is constructed of 14 in (360 mm) thick solid steel and weighs in excess of 50 tons.

Intermediate-level waste (ILW) contains higher amounts of radioactivity compared to low-level waste. It generally requires shielding, but not cooling. Intermediate-level wastes includes resins, chemical sludge and metal nuclear fuel cladding, as well as contaminated materials from reactor decommissioning. It may be solidified in concrete or bitumen or mixed with silica sand and vitrified for disposal. As a general rule, short-lived waste (mainly non-fuel materials from reactors) is buried in shallow repositories, while long-lived waste (from fuel and fuel reprocessing) is deposited in geological repository. U.S. regulations do not define this category of waste; the term is used in Europe and elsewhere.

High-level Waste

High-level waste (HLW) is produced by nuclear reactors. The exact definition of HLW differs internationally. After a nuclear fuel rod serves one fuel cycle and is removed from the core, it is considered HLW. Fuel rods contain fission products and transuranic elements generated in the reactor core. Spent fuel is highly radioactive and often hot. HLW accounts for over 95 percent of the total radioactivity produced in the process of nuclear electricity generation.

The radioactive waste from spent fuel rods consist primarily of cesium-137 and strontium-90, but it may also include plutonium, which can be considered a transuranic waste. The half-lives of these radioactive elements can differ quite extremely. Some elements, such as cesium-137 and strontium-90 have half-lives of approximately 30 years. Meanwhile, plutonium has a half-life of that can stretch to as long as 24,000 years.

The amount of HLW worldwide is currently increasing by about 12,000 metric tons every year. A 1000-megawatt nuclear power plant produces about 27 tonnes of spent nuclear fuel (unreprocessed) every year.

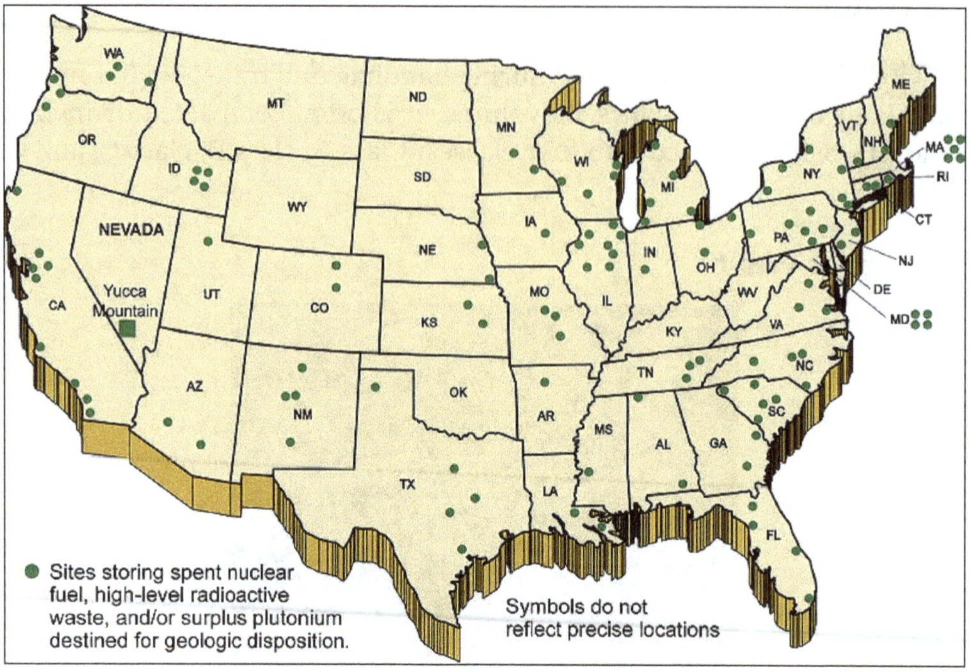

The current locations across the U.S. where nuclear waste is stored.

In 2010, it was estimated that about 250,000 tons of nuclear HLW were stored globally. This does not include amounts that have escaped into the environment from accidents or tests. Japan is estimated to hold 17,000 tons of HLW in storage in 2015. As of 2019, the United States has over 90,000 tons of HLW. HLW have been shipped to other countries to be stored or reprocessed, and in some cases, shipped back as active fuel.

The ongoing controversy over high-level radioactive waste disposal is a major constraint on the nuclear power's global expansion. Most scientists agree that the main proposed long-term solution is deep geological burial, either in a mine or a deep borehole. However, as of 2019, no government has succeeded in opening such a repository for civilian high-level nuclear waste. Finland is in the advanced stage of the construction of such facility, the Onkalo spent nuclear fuel repository. The Morris Operation is currently the only de facto high-level radioactive waste storage site in the United States.

Reprocessing or recycling spent nuclear fuel options already available or under active development still generate some waste and are therefore not a total solution, but can significantly reduce the quantity of waste. There are many active reprocessing programs worldwide.

Transuranic Waste

Transuranic waste (TRUW) as defined by U.S. regulations is, without regard to form or origin, waste that is contaminated with alpha-emitting transuranic radionuclides with half-lives greater than 20 years and concentrations greater than 100 nCi/g (3.7 MBq/kg), excluding high-level waste. Elements that have an atomic number greater than uranium are called transuranic ("beyond uranium"). Because of their long half-lives, TRUW is disposed more cautiously than either low- or intermediate-level waste. In the U.S., it arises mainly from nuclear weapons production, and consists of clothing, tools, rags, residues, debris and other items contaminated with small amounts of radioactive elements (mainly plutonium).

Under U.S. law, transuranic waste is further categorized into "contact-handled" (CH) and "remote-handled" (RH) on the basis of the radiation dose rate measured at the surface of the waste container. CH TRUW has a surface dose rate not greater than 200 mrem per hour (2 mSv/h), whereas RH TRUW has a surface dose rate of 200 mrem/h (2 mSv/h) or greater. CH TRUW does not have the very high radioactivity of high-level waste, nor its high heat generation, but RH TRUW can be highly radioactive, with surface dose rates up to 1,000,000 mrem/h (10,000 mSv/h). The U.S. currently disposes of TRUW generated from military facilities at the Waste Isolation Pilot Plant (WIPP) in a deep salt formation in New Mexico.

Prevention

A theoretical way to reduce waste accumulation is to phase out current reactors in favour of Generation IV reactors, which output less waste per power generated. Fast reactors can theoretically consume some existing waste. The UK's Nuclear Decommissioning Authority published a position paper in 2014 on the progress on approaches to the management of separated plutonium, which summarises the conclusions of the work that NDA shared with UK government.

HOUSEHOLD HAZARDOUS WASTE

Household hazardous waste (HHW), sometimes called retail hazardous waste or "home generated special materials', is post-consumer waste which qualifies as hazardous waste when discarded. It includes household chemicals and other substances for which the owner no longer has a use, such as consumer products sold for home care, personal care, automotive care, pest control and other purposes. These products exhibit many of the same dangerous characteristics as fully regulated hazardous waste due to their potential for reactivity, ignitability, corrosivity, toxicity, or persistence. Examples include drain cleaners, oil paint, motor oil, antifreeze, fuel, poisons, pesticides, herbicides and rodenticides, fluorescent lamps, lamp ballasts, smoke detectors, medical waste, some types of cleaning chemicals, and consumer electronics (such as televisions, computers, and cell phones).

Certain items such as batteries and fluorescent lamps can be returned to retail stores for disposal. The Rechargeable Battery Recycling Corporation (RBRC) maintains a list of battery recycling locations and your local environmental organization should have list of fluorescent lamp recycling locations. The classification "household hazardous waste" has been used for decades and does not accurately reflect the larger group of materials that during the past several years have become known as "household hazardous wastes". These include items such as latex paint, non-hazardous household products and other items that do not generally exhibit hazardous characteristics which are routinely included in "household hazardous waste" disposal programs. The term "home generated special materials" more accurately identifies a broader range of items that public agencies are targeting as recyclable and should not be disposed of into a landfill.

HHW is not regulated by the EPA. Many states and local solid waste management departments have created and funded Household Hazardous Waste collection programs to offer safe disposal options. These programs may include home collection service, permanent facilities and one day collection events.

Although most U.S. states and federal regulations continue to permit homeowner disposal of some household hazardous waste into the solid waste stream, state agencies are becoming more stringent in enforcing existing hazardous waste regulations at the retail level.

The professional organization most focused on HHW issues is the North American Hazardous Materials Management Association, NAHMMA. NAHMMA has chapters in many states, holds an annual conference, provides training and offers professional publications. In collaboration with the Solid Waste Association of North America (SWANA) NAHMMA offers certification to HHW collection professionals.

Chemical Waste

Chemical waste is typically generated by factories, processing centers, warehouses, and plants. This waste may Different Kinds of Waste Atlanta include harmful or dangerous chemicals and chemical residue, and waste disposal must adhere to careful guidelines.

- Collected as a Hazardous Waste: A chemical which exhibits a 'hazardous' characteristic, is listed according to Federal or State regulation, or best management practices dictate strict control must be managed as a hazardous waste. Most of the information in this document describes the rules to follow when managing hazardous wastes.
    - Hazardous wastes have the greatest potential to harm people or the environment, and bear the full burden of environmental regulation.
    - Mismanagement of hazardous wastes can result in significant environmental penalties and enforcement actions.
    - The term 'hazardous waste' should only be used when describing chemicals. Infectious or radioactive wastes are not 'hazardous wastes' unless they are mixed with chemical wastes.
    - Hazardous wastes are generated routinely.
    - Collection and management of chemical wastes as hazardous wastes ensures the maximum level of environmental protection and is the safest means of chemical waste management.
- Collected as a Non-hazardous Waste: A chemical which does not exhibit a state or federal hazardous characteristic and isn't listed as a 'hazardous waste' isn't necessarily safe for disposal via sink or in the regular trash.
    - In some cases, protection of the environment requires us to go above and beyond the hazardous waste regulations. An example would be ethidium bromide, which doesn't technically exhibit a hazardous characteristic but which isn't safe for handling as regular trash. Other examples include materials like nanoparticles for which comprehensive safety information does not yet exist. Often the precautionary principle dictates that we collect these materials because we do not understand the potential harm they can cause in the environment.
    - In other cases, a different regulation might stipulate collection of a waste chemical. An

example would be very small concentrations (part-per-billion level) of mercury in otherwise benign reagents. Local Clean Water Act enforcement prohibits mercury disposal to sink drains even though the level of contamination is low.

  - In cases where a chemical is collected as a non-hazardous waste, some of the hazardous waste management rules do not apply.

- Collected as a Universal Waste: A small subset of chemical wastes have been de-regulated to some extent based on the fact that they are so widespread.
  - Management regulations still exist, but they are lessened relative to hazardous waste regulations.
  - Universal wastes are most often generated and managed through Facilities Management Department operations.
  - Universal Wastes include:
    - Fluorescent bulbs of all shapes and sizes.
    - Many types of batteries.
    - Mercury-containing devices such as switches and thermostats.
- Safe for Sink or Trash disposal: A very small percentage of chemical wastes are un-regulated and safe to pour into sinks or place in the trash. Some examples include benign salts like sodium chloride, and non-toxic, non-corrosive cleaning chemicals. Contact EHS prior to disposing any chemical into a sink or via the trash.

Hazardous Wastes

The most important question to answer when managing a chemical waste: "Does my chemical waste have to be collected and managed as a hazardous waste?" It is necessary to know the answer to this question as this is the first step in a process known as making a 'waste determination', and is a required step in chemical waste management.

In order to determine whether a chemical waste must be collected as a hazardous waste you must have information about the properties and behavior of the chemical. This knowledge is fundamental to safe laboratory operations, and literature is available in your workplace in the form of Safety Data Sheets (SDS or MSDS) which can help. If you don't know or are unsure about the characteristics or properties of a chemical waste contact EHS for assistance.

Never assume a chemical waste is safe for disposal in the regular trash or via a sink or drain. Always err on the side of caution; collection and management as a hazardous waste because this is the safest and most responsible way to collect a chemical waste.

A chemical waste must be classified and managed as a hazardous waste if it exhibits any of the four characteristics described below, or is specifically listed in the regulations.

Ignitable Characteristic

A chemical waste is a hazardous waste due to ignitability if:

- Liquid Chemicals: The flash point of the liquid chemical is less than or equal to 140 degrees F.
    - Equivalent is 60 degrees C.
    - Flashpoint determination is 'Pensky-Martens Closed Cup Tester'.
    - Common examples include:
        - Alcohols (For ethanol, mixtures greater than or equal to 20% are hazardous wastes. For other alcohols the cut-off is 10%).
        - Organic Solvents and mixtures containing organic solvents such as xylenes, hexane, toluene, acetone, etc.
        - Stains and mixtures containing stains (because they are solvent-based).
        - Oil-based paints and coatings.
- Solid Chemicals: The chemical is capable, under standard temperature and pressure, of causing fire through friction, absorption of moisture or spontaneous chemical changes, and burns vigorously when ignited.
    - Common examples include:
        - Paraformaldehyde.
        - Parafin wax with xylene.
        - Rags saturated with an ignitable liquid.
- Compressed Gas: Ignitable compressed gases must also be managed as hazardous wastes.
    - Generally this involves partially-full, or left-over cylinders of gas.
    - Common examples include:
        - Hydrogen.
        - Acetylene.
        - Propane.
        - Butane.
- Oxidizers: The chemical is capable of enhancing the combustion of other materials, generally by yielding oxygen.
    - Common examples include:
        - Chlorates.

        - Chlorites.
        - Nitrates.
        - Perchlorates.
        - Perchlorites.
        - Permanganates.
        - Peroxides.

If your chemical waste exhibits any of the 'ignitable' characteristics above, you must manage it as an ignitable hazardous waste.

Corrosive Characteristic

A chemical waste is a hazardous waste due to corrosivity if:

- It is aqueous and has a pH less than or equal to 2, or greater than or equal to 12.5.
- It is a liquid and corrodes steel (Type SAE 1020) at a rate greater than 6.35 mm (approximately 0.250 inch) per year.
- Common examples include:
    - Hydrochloric Acid.
    - Sulfuric Acid.
    - Nitric Acid.
    - Sodium Hydroxide.

Reactive Characteristic

A reactive hazardous waste is defined as a material which:

- Under normal conditions is unstable and can undergo violent changes without detonating.
- Reacts violently with Water.
    - Common examples include:
        - Sodium metal.
        - Anhydrides.
        - Sodium Borohydride.
- Reacts violently with Air.
    - Common Examples Include:
        - tert-butyllithium.

        - Capable of detonation or violent explosion.
    - Common examples include:
        - Dry picric acid.
        - Azide compounds.
        - Organic peroxides.
        - Old ether or tetrahydrofuran with peroxide formation.
- A cyanide or sulfide which, when exposed to a pH of between 2 and 12.5, generates toxic gases, vapors or fumes.
    - Common examples include:
        - Sodium cyanide.
        - Potassium cyanide.
        - Sodium sulfide.
        - Carbon disulfide.

Toxic

The toxic 'characteristic' is where the regulations start to get into listing specific chemicals. To determine whether a chemical waste exhibits the toxic characteristic, it is necessary check the federal toxic list, known as the 'D' list. The state of Massachusetts adds some chemicals to the list as well.

As a general rule, a waste that contains any material on this list should be collected for disposal regardless of concentration; even if it's not technically regulated as a hazardous waste it still doesn't belong in the environment.

FLY ASH

Fly ash or flue ash, also known as pulverised fuel ash in the United Kingdom, is a coal combustion product that is composed of the particulates (fine particles of burned fuel) that are driven out of coal-fired boilers together with the flue gases. Ash that falls to the bottom of the boiler's combustion chamber (commonly called a firebox) is called bottom ash. In modern coal-fired power plants, fly ash is generally captured by electrostatic precipitators or other particle filtration equipment before the flue gases reach the chimneys. Together with bottom ash removed from the bottom of the boiler, it is known as coal ash. Depending upon the source and composition of the coal being burned, the components of fly ash vary considerably, but all fly ash includes substantial amounts of silicon dioxide (SiO_2) (both amorphous and crystalline), aluminium oxide (Al_2O_3) and calcium oxide (CaO), the main mineral compounds in coal-bearing rock strata.

The minor constituents of fly ash depend upon the specific coal bed composition but may include one or more of the following elements or compounds found in trace concentrations (up to hundreds ppm): Arsenic, beryllium, boron, cadmium, chromium, hexavalent chromium, cobalt, lead, manganese, mercury, molybdenum, selenium, strontium, thallium, and vanadium, along with very small concentrations of dioxins and PAH compounds. It also has unburnt carbon.

In the past, fly ash was generally released into the atmosphere, but air pollution control standards now require that it be captured prior to release by fitting pollution control equipment. In the United States, fly ash is generally stored at coal power plants or placed in landfills. About 43% is recycled, often used as a pozzolan to produce hydraulic cement or hydraulic plaster and a replacement or partial replacement for Portland cement in concrete production. Pozzolans ensure the setting of concrete and plaster and provide concrete with more protection from wet conditions and chemical attack.

In the case that fly (or bottom) ash is not produced from coal, for example when solid waste is incinerated in a waste-to-energy facility to produce electricity, the ash may contain higher levels of contaminants than coal ash. In that case the ash produced is often classified as hazardous waste.

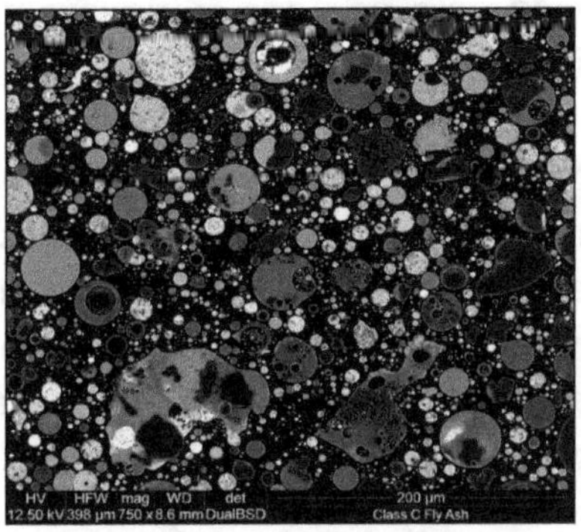

Photomicrograph made with a scanning electron microscope (SEM) and back-scatter detector: Cross section of fly ash particles at 750x magnification.

Chemical Composition and Classification

Component	Bituminous	Subbituminous	Lignite
SiO_2 (%)	20-60	40-60	15-45
Al_2O_3 (%)	5-35	20-30	20-25
Fe_2O_3 (%)	10-40	4-10	4-15
CaO (%)	1-12	5-30	15-40
LOI (%)	0-15	0-3	0-5

Fly ash material solidifies while suspended in the exhaust gases and is collected by electrostatic precipitators or filter bags. Since the particles solidify rapidly while suspended in the exhaust gases, fly ash particles are generally spherical in shape and range in size from 0.5 µm to 300 µm. The major consequence of the rapid cooling is that few minerals have time to crystallize, and that mainly amorphous, quenched glass remains. Nevertheless, some refractory phases in the pulverized coal do not melt (entirely), and remain crystalline. In consequence, fly ash is a heterogeneous material. SiO_2, Al_2O_3, Fe_2O_3 and occasionally CaO are the main chemical components present in fly ashes. The mineralogy of fly ashes is very diverse. The main phases encountered are a glass phase, together with quartz, mullite and the iron oxides hematite, magnetite and maghemite. Other phases often identified are cristobalite, anhydrite, free lime, periclase, calcite, sylvite, halite, portlandite, rutile and anatase. The Ca-bearing minerals anorthite, gehlenite, akermanite and various calcium silicates and calcium aluminates identical to those found in Portland cement can be identified in Ca-rich fly ashes. The mercury content can reach 1 ppm, but is generally included in the range 0.01 - 1 ppm for bituminous coal. The concentrations of other trace elements vary as well according to the kind of coal combusted to form it. In fact, in the case of bituminous coal, with the notable exception of boron, trace element concentrations are generally similar to trace element concentrations in unpolluted soils.

Two classes of fly ash are defined by ASTM C618: Class F fly ash and Class C fly ash. The chief difference between these classes is the amount of calcium, silica, alumina, and iron content in the ash. The chemical properties of the fly ash are largely influenced by the chemical content of the coal burned (i.e., anthracite, bituminous, and lignite).

Not all fly ashes meet ASTM C618 requirements, although depending on the application, this may not be necessary. Fly ash used as a cement replacement must meet strict construction standards, but no standard environmental regulations have been established in the United States. Seventy-five percent of the fly ash must have a fineness of 45 µm or less, and have a carbon content, measured by the loss on ignition (LOI), of less than 4%. In the US, LOI must be under 6%. The particle size distribution of raw fly ash tends to fluctuate constantly, due to changing performance of the coal mills and the boiler performance. This makes it necessary that, if fly ash is used in an optimal way to replace cement in concrete production, it must be processed using beneficiation methods like mechanical air classification. But if fly ash is used as a filler to replace sand in concrete production, unbeneficiated fly ash with higher LOI can be also used. Especially important is the ongoing quality verification. This is mainly expressed by quality control seals like the Bureau of Indian Standards mark or the DCL mark of the Dubai Municipality.

Class F

The burning of harder, older anthracite and bituminous coal typically produces Class F fly ash. This fly ash is pozzolanic in nature, and contains less than 7% lime (CaO). Possessing pozzolanic properties, the glassy silica and alumina of Class F fly ash requires a cementing agent, such as Portland cement, quicklime, or hydrated lime—mixed with water to react and produce cementitious compounds. Alternatively, adding a chemical activator such as sodium silicate (water glass) to a Class F ash can form a geopolymer.

Class C

Fly ash produced from the burning of younger lignite or sub-bituminous coal, in addition to having pozzolanic properties, also has some self-cementing properties. In the presence of water, Class C fly ash hardens and gets stronger over time. Class C fly ash generally contains more than 20% lime (CaO). Unlike Class F, self-cementing Class C fly ash does not require an activator. Alkali and sulfate (SO_4) contents are generally higher in Class C fly ashes.

At least one US manufacturer has announced a fly ash brick containing up to 50% Class C fly ash. Testing shows the bricks meet or exceed the performance standards listed in ASTM C 216 for conventional clay brick. It is also within the allowable shrinkage limits for concrete brick in ASTM C 55, Standard Specification for Concrete Building Brick. It is estimated that the production method used in fly ash bricks will reduce the embodied energy of masonry construction by up to 90%. Bricks and pavers were expected to be available in commercial quantities before the end of 2009.

Disposal and Market Sources

In the past, fly ash produced from coal combustion was simply entrained in flue gases and dispersed into the atmosphere. This created environmental and health concerns that prompted laws that have reduced fly ash emissions to less than 1% of ash produced. Worldwide, more than 65% of fly ash produced from coal power stations is disposed of in landfills and ash ponds.

Ash that is stored or deposited outdoors can eventually leach toxic compounds into underground water aquifers. For this reason, much of the current debate around fly ash disposal revolves around creating specially lined landfills that prevent the chemical compounds from being leached into the ground water and local ecosystems. Since coal was the dominant energy source in the United States for many decades, power companies often located their coal plants nearby metropolitan areas. Compounding the environmental issues, the coal plants need significant amounts of water to operate their boilers, leading coal plants (and later their fly ash storage basins) to be located near metropolitan areas and near rivers and lakes which are often used as drinking supplies by nearby cities. Many of those fly ash basins were unlined and also were are great risk of spilling and flooding from nearby rivers and lakes. For example, Duke Energy in North Carolina has been involved in several major lawsuits related to its coal ash storage and spills into the leakage of ash into the water basin.

The recycling of fly ash has become an increasing concern in recent years due to increasing landfill costs and current interest in sustainable development. As of 2017, coal-fired power plants in the US reported producing 38.2 million short tons (34.7×10^6 t) of fly ash, of which 24.1 million short tons (21.9×10^6 t) were reused in various applications. Environmental benefits to recycling fly ash includes reducing the demand for virgin materials that would need quarrying and cheap substitution for materials such as Portland cement.

Reuse

There is no US governmental registration or labelling of fly ash utilization in the different sectors of the economy - industry, infrastructures and agriculture. Fly ash utilization survey data, acknowledged as incomplete, are published annually by the American Coal Ash Association.

Coal ash uses include (approximately in order of decreasing importance):

- Concrete production, as a substitute material for Portland cement,sand.
- Fly-ash pellets which can replace normal aggregate in concrete mixture.
- Embankments and other structural fills (usually for road construction).
- Grout and flowable fill production.

- Waste stabilization and solidification.
- Cement clinker production - (as a substitute material for clay).
- Mine reclamation.
- Stabilization of soft soils.
- Road subbase construction.
- As aggregate substitute material (e.g. for brick production).
- Mineral filler in asphaltic concrete.
- Agricultural uses: Soil amendment, fertilizer, cattle feeders, soil stabilization in stock feed yards, and agricultural stakes.
- Loose application on rivers to melt ice.
- Loose application on roads and parking lots for ice control.

Other applications include cosmetics, toothpaste, kitchen counter tops, floor and ceiling tiles, bowling balls, flotation devices, stucco, utensils, tool handles, picture frames, auto bodies and boat hulls, cellular concrete, geopolymers, roofing tiles, roofing granules, decking, fireplace mantles, cinder block, PVC pipe, structural insulated panels, house siding and trim, running tracks, blasting grit, recycled plastic lumber, utility poles and crossarms, railway sleepers, highway noise barriers, marine pilings, doors, window frames, scaffolding, sign posts, crypts, columns, railroad ties, vinyl flooring, paving stones, shower stalls, garage doors, park benches, landscape timbers, planters, pallet blocks, molding, mail boxes, artificial reef, binding agent, paints and undercoatings, metal castings, and filler in wood and plastic products.

Portland Cement

Owing to its pozzolanic properties, fly ash is used as a replacement for Portland cement in concrete. The use of fly ash as a pozzolanic ingredient was recognized as early as 1914, although the earliest noteworthy study of its use was in 1937. Roman structures such as aqueducts or the Pantheon in Rome used volcanic ash or pozzolana (which possesses similar properties to fly ash) as pozzolan in their concrete. As pozzolan greatly improves the strength and durability of concrete, the use of ash is a key factor in their preservation.

Use of fly ash as a partial replacement for Portland cement is particularly suitable but not limited to Class C fly ashes. Class "F" fly ashes can have volatile effects on the entrained air content of concrete, causing reduced resistance to freeze/thaw damage. Fly ash often replaces up to 30% by mass of Portland cement, but can be used in higher dosages in certain applications. In some cases, fly ash can add to the concrete's final strength and increase its chemical resistance and durability.

Fly ash can significantly improve the workability of concrete. Recently, techniques have been developed to replace partial cement with high-volume fly ash (50% cement replacement). For roller-compacted concrete (RCC) [used in dam construction], replacement values of 70% have been achieved with processed fly ash at the Ghatghar dam project in Maharashtra, India. Due to the spherical shape of fly ash particles, it can increase workability of cement while reducing water demand. Proponents of fly ash claim that replacing Portland cement with fly ash reduces the greenhouse gas "footprint" of concrete, as the production of one ton of Portland cement generates approximately one ton of CO_2, compared to no CO_2 generated with fly ash. New fly ash production, i.e., the burning of coal, produces approximately 20 to 30 tons of CO_2 per ton of fly ash. Since the worldwide production of Portland cement is expected to reach nearly 2 billion tons by 2010, replacement of any large portion of this cement by fly ash could significantly reduce carbon emissions associated with construction, as long as the comparison takes the production of fly ash as a given.

Embankment

Fly ash properties are unusual among engineering materials. Unlike soils typically used for embankment construction, fly ash has a large uniformity coefficient and it consists of clay-sized particles. Engineering properties that affect the use of fly ash in embankments include grain size distribution, compaction characteristics, shear strength, compressibility, permeability, and frost susceptibility. Nearly all the types of fly ash used in embankments are Class F.

Soil Stabilization

Soil stabilization is the permanent physical and chemical alteration of soils to enhance their physical properties. Stabilization can increase the shear strength of a soil and control the shrink-swell properties of a soil, thus improving the load-bearing capacity of a sub-grade to support pavements and foundations. S3.tabilization can be used to treat a wide range of sub-grade materials from expansive clays to granular materials. Stabilization can be achieved with a variety of chemical additives including lime, fly ash, and Portland cement. Proper design and testing is an important component of any stabilization project. This allows for the establishment of design criteria, and determination of the proper chemical additive and admixture rate that achieves the desired engineering properties. Stabilization process benefits can include: Higher resistance (R) values, Reduction in plasticity, Lower permeability, Reduction of pavement thickness, Elimination of excavation - material hauling/handling - and base importation, Aids compaction, Provides "all-weather" access onto and within projects sites. Another form of soil treatment closely related to soil stabilization is soil modification, sometimes referred to as "mud drying" or soil conditioning. Although some stabilization inherently occurs in soil modification, the distinction is that soil modification is merely a means to reduce the moisture content of a soil to expedite construction, whereas stabilization can substantially increase the shear strength of a material such that it can be incorporated into the project's structural design. The determining factors associated with soil modification vs soil stabilization may be the existing moisture content, the end use of the soil structure and ultimately the cost benefit provided. Equipment for the stabilization and modification processes include: chemical additive spreaders, soil mixers (reclaimers), portable pneumatic storage containers, water trucks, deep lift compactors, motor graders.

Flowable Fill

Fly ash is also used as a component in the production of flowable fill (also called controlled low strength material, or CLSM), which is used as self-leveling, self-compact backfill material in lieu of compacted earth or granular fill. The strength of flowable fill mixes can range from 50 to 1,200 lbf/in^2 (0.3 to 8.3 MPa), depending on the design requirements of the project in question. Flowable fill includes mixtures of Portland cement and filler material, and can contain mineral admixtures. Fly ash can replace either the Portland cement or fine aggregate (in most cases, river sand) as a filler material. High fly ash content mixes contain nearly all fly ash, with a small percentage of Portland cement and enough water to make the mix flowable. Low fly ash content mixes contain a high percentage of filler material, and a low percentage of fly ash, Portland cement, and water. Class F fly ash is best suited for high fly ash content mixes, whereas Class C fly ash is almost always used in low fly ash content mixes.

Asphalt Concrete

Asphalt concrete is a composite material consisting of an asphalt binder and mineral aggregate commonly used to surface roads. Both Class F and Class C fly ash can typically be used as a mineral filler to fill the voids and provide contact points between larger aggregate particles in asphalt concrete mixes. This application is used in conjunction,or as a replacement for, other binders (such as Portland cement or hydrated lime). For use in asphalt pavement, the fly ash must meet mineral filler specifications outlined in ASTM D242. The hydrophobic nature of fly ash gives pavements better resistance to stripping. Fly ash has also been shown to increase the stiffness of the asphalt matrix, improving rutting resistance and increasing mix durability.

Geopolymers

More recently, fly ash has been used as a component in geopolymers, where the reactivity of the fly ash glasses can be used to create a binder similar to a hydrated Portland cement in appearance, but with potentially superior properties, including reduced CO_2 emissions, depending on the formulation.

Roller Compacted Concrete

The upper reservoir of Ameren's Taum Sauk hydroelectric plant was constructed of roller-compacted concrete that included fly ash from one of Ameren's coal plants.

Another application of using fly ash is in roller compacted concrete dams. Many dams in the US have been constructed with high fly ash contents. Fly ash lowers the heat of hydration allowing thicker placements to occur. Data for these can be found at the US Bureau of Reclamation.

Bricks

There are several techniques for manufacturing construction bricks from fly ash, producing a wide variety of products. One type of fly ash brick is manufactured by mixing fly ash with an equal amount of clay, then firing in a kiln at about 1000 °C. This approach has the principal benefit of reducing the amount of clay required. Another type of fly ash brick is made by mixing soil, plaster of paris, fly ash and water, and allowing the mixture to dry. Because no heat is required, this technique reduces air pollution. More modern manufacturing processes use a greater proportion of fly ash, and a high pressure manufacturing technique, which produces high strength bricks with environmental benefits.

In the United Kingdom, fly ash has been used for over fifty years to make concrete building blocks. They are widely used for the inner skin of cavity walls. They are naturally more thermally insulating than blocks made with other aggregates.

Ash bricks have been used in house construction in Windhoek, Namibia since the 1970s. There is, however, a problem with the bricks in that they tend to fail or produce unsightly pop-outs. This happens when the bricks come into contact with moisture and a chemical reaction occurs causing the bricks to expand.

Leading manufacturers use an industrial standard known as "Pulverized fuel ash for lime-Pozzolana mixture" using over 75% post-industrial recycled waste, and a compression process. This produces a strong product with good insulation properties and environmental benefits.

Metal Matrix Composites

Fly ash particles have proved its potential as good reinforcement with aluminum alloys and show the improvement of physical and mechanical properties; in particular, the compression strength, tensile strength, and hardness increase when the percentage of fly ash content is increased, whereas the density decreases. The presence of fly ash cenospheres in pure Al matrix decreases its CTE(Coefficient of Thermal Expansion).

Waste Treatment and Stabilization

Fly ash, in view of its alkalinity and water absorption capacity, may be used in combination with other alkaline materials to transform sewage sludge into organic fertilizer or biofuel.

Catalyst

Fly ash, when treated with sodium hydroxide, appears to function well as a catalyst for converting polyethylene into substance similar to crude oil in a high-temperature process called pyrolysis.

In addition, fly ash, mainly class C, may be used in the stabilization/solidification process of hazardous wastes and contaminated soils. For example, the Rhenipal process uses fly ash as an admixture to stabilize sewage sludge and other toxic sludges. This process has been used since 1996 to stabilize large amounts of chromium(VI) contaminated leather sludges in Alcanena, Portugal.

Environmental Problems

Groundwater Contamination

Coal contains trace levels of trace elements (such as arsenic, barium, beryllium, boron, cadmium, chromium, thallium, selenium, molybdenum and mercury), many of which are highly toxic to humans and other life. Therefore, fly ash obtained after combustion of this coal contains enhanced concentrations of these elements and the potential of the ash to cause groundwater pollution is significant. In the USA there are documented cases of groundwater pollution that followed ash disposal or utilization without the necessary protection having been put in place. For example, a December 2008 Maryland court decision levied a $54 million penalty against Constellation Energy, which had performed a “restoration project” of filling an abandoned gravel quarry with fly ash; the ash contaminated area waterwells with heavy metals.

Examples of Groundwater Contamination

In 2014, residents living near the Buck Steam Station in Dukeville, North Carolina, were told that “coal ash pits near their homes could be leaching dangerous materials into groundwater.”

Illinois has many coal ash dumpsites with coal ash generated by coal-burning electric power plants. Of the state’s 24 coal ash dumpsites with available data, 22 have released toxic pollutants including arsenic, cobalt, and lithium, into groundwater, rivers and lakes. The hazardous toxic chemicals dumped into the water in Illinois by these coal ash dumpsites include more than 300,000 pounds of aluminum, 600 pounds of arsenic, nearly 300,000 pounds of boron, over 200 pounds of cadmium, over 15,000 pounds of manganese, roughly 1,500 pounds of selenium, roughly 500,000 pounds of nitrogen, and nearly 40 million pounds of sulfate, according to a report by the Environmental Integrity Project, Earthjustice, the Prairie Rivers Network, and the Sierra Club.

Groundwater surrounding every single one of the 16 coal-burning power plants in Texas has been polluted by coal ash, according to a study by the Environmental Integrity Project (EIP). Unsafe levels of arsenic, cobalt, lithium, and other contaminants were found in the groundwater near all the ash dump sites. At 12 of the 16 sites, the EIP analysis found levels of arsenic in the groundwater 10 times higher than the EPA Maximum Contaminant Level; arsenic has been found to cause several types of cancer. At 10 of the sites, lithium, which causes neurological disease, was found in the groundwater at concentrations more than 1,000 micrograms per liter, which is 25 times the maximum acceptable level. The report concludes that the fossil fuel industry in Texas has failed to comply with federal regulations on coal ash processing, and state regulators have failed to protect the groundwater.

Ecology

The effect of fly ash on the environment can vary based on the thermal power plant where it is produced, as well as the proportion of fly ash to bottom ash in the waste product. This is due to the different chemical make-up of the coal based on the geology of the area the coal is found and the burning process of the coal in the power plant. When the coal is combusted, it creates an alkaline dust. This alkaline dust can have a pH ranging from 8 to as high as 12. Fly ash dust can be deposited on topsoil increasing the pH and affecting the plants and animals in the surrounding ecosystem. Trace elements, such as, iron, manganese, zinc, copper, lead, nickel, chromium, cobalt, arsenic, cadmium, and mercury, can be found at higher concentrations compared to bottom ash and the parent coal.

Fly ash can leach toxic constituents that can be anywhere from one hundred to one thousand times greater than the federal standard for drinking water. Fly ash can contaminate surface water through erosion, surface runoff, airborne particles landing on the water surface, contaminated ground water moving into surface waters, flooding drainage, or discharge from a coal ash pond. Fish can be contaminated a couple of different ways. When the water is contaminated by fly ash, the fish can absorb the toxins through their gills. The sediment in the water can also become contaminated. The contaminated sediment can contaminate the food sources for the fish, the fish can then become contaminated from consuming those food sources. This can then lead to contamination of organisms that consume these fish, such as, birds, bear, and even humans. Once exposed to fly ash contaminating the water, aquatic organisms have had increased levels of calcium, zinc, bromine, gold, cerium, chromium, selenium, cadmium, and mercury.

Soils contaminated by fly ash showed an increase in bulk density and water capacity, but a decrease in hydraulic conductivity and cohesiveness. The effect of fly ash on soils and microorganisms in the soils are influenced by the pH of the ash and trace metal concentrations in the ash. Microbial communities in contaminated soil have shown reductions in respiration and nitrification. These contaminated soils can be detrimental or beneficial to plant development. Fly ash typically has beneficial outcomes when it corrects nutrient deficiencies in the soil. Most detrimental effects were observed when boron phytotoxicity was observed. Plants absorb elements elevated by the fly ash from the soil. Arsenic, molybdenum, and selenium were the only elements found at potentially toxic levels for grazing animals. Terrestrial organisms exposed to fly ash only showed increased levels of selenium.

Spills of Bulk Storage

Tennessee Valley Authority Fly Ash containment failure.

Where fly ash is stored in bulk, it is usually stored wet rather than dry to minimize fugitive dust. The resulting impoundments (ponds) are typically large and stable for long periods, but any breach of their dams or bunding is rapid and on a massive scale.

In December 2008, the collapse of an embankment at an impoundment for wet storage of fly ash at the Tennessee Valley Authority's Kingston Fossil Plant caused a major release of 5.4 million cubic yards of coal fly ash, damaging 3 homes and flowing into the Emory River. Cleanup costs may exceed \$1.2 billion. This spill was followed a few weeks later by a smaller TVA-plant spill in Alabama, which contaminated Widows Creek and the Tennessee River.

In 2014, 39,000 tons of ash and 27 million gallons (100,000 cubic meters) of contaminated water spilled into the Dan River near Eden, NC from a closed North Carolina coal-fired power plant that is owned by Duke Energy. It is currently the third worst coal ash spill ever to happen in the United States.

The U.S. Environmental Protection Agency (EPA) published a Coal Combustion Residuals (CCR) regulation in 2015. The agency continued to classify coal ash as non-hazardous (thereby avoiding strict permitting requirements under Subtitle C of the Resource Conservation and Recovery Act (RCRA), but with new restrictions:

- Existing ash ponds that are contaminating groundwater must stop receiving CCR, and close or retrofit with a liner.
- Existing ash ponds and landfills must comply with structural and location restrictions, where applicable, or close.
- A pond no longer receiving CCR is still subject to all regulations unless it is dewatered and covered by 2018.
- New ponds and landfills must include a geomembrane liner over a layer of compacted soil.

The regulation was designed to prevent pond failures and protect groundwater. Enhanced inspection, record keeping and monitoring is required. Procedures for closure are also included and include capping, liners, and dewatering. The CCR regulation has since been subject to litigation.

Contaminants

Fly ash contains trace concentrations of heavy metals and other substances that are known to be detrimental to health in sufficient quantities. Potentially toxic trace elements in coal include arsenic, beryllium, cadmium, barium, chromium, copper, lead, mercury, molybdenum, nickel, radium, selenium, thorium, uranium, vanadium, and zinc. Approximately 10% of the mass of coals burned in the United States consists of unburnable mineral material that becomes ash, so the concentration of most trace elements in coal ash is approximately 10 times the concentration in the original coal. A 1997 analysis by the United States Geological Survey (USGS) found that fly ash typically contained 10 to 30 ppm of uranium, comparable to the levels found in some granitic rocks, phosphate rock, and black shale.

In 1980 the U.S. Congress defined coal ash as a "special waste" that would not be regulated under the stringent hazardous waste permitting requirements of RCRA. In its amendments to RCRA, Congress directed EPA to study the special waste issue and make a determination as to whether stricter permit regulation was necessary. In 2000, EPA stated that coal fly ash did not need to be regulated as a hazardous waste. As a result, most power plants were not required to install geomembranes or leachate collection systems in ash ponds.

Studies by the USGS and others of radioactive elements in coal ash have concluded that fly ash compares with common soils or rocks and should not be the source of alarm. However, community and environmental organizations have documented numerous environmental contamination and damage concerns.

Exposure Concerns

Crystalline silica and lime along with toxic chemicals represent exposure risks to human health and the environment. Fly ash contains crystalline silica which is known to cause lung disease, in particular silicosis, if inhaled. Crystalline silica is listed by the IARC and US National Toxicology Program as a known human carcinogen.

Lime (CaO) reacts with water (H_2O) to form calcium hydroxide [$Ca(OH)_2$], giving fly ash a pH somewhere between 10 and 12, a medium to strong base. This can also cause lung damage if present in sufficient quantities.

Material Safety Data Sheets recommend a number of safety precautions be taken when handling or working with fly ash. These include wearing protective goggles, respirators and disposable clothing and avoiding agitating the fly ash in order to minimize the amount which becomes airborne.

The National Academy of Sciences noted in 2007 that "the presence of high contaminant levels in many CCR (coal combustion residue) leachates may create human health and ecological concerns".

BRINE WASTE

In general, "brine" is any solution with an extremely high concentration of salts such as sodium chloride, which can occur either naturally (as with seawater, deep-water ocean pools, salt lakes, etc.) or as a byproduct of industry. These byproducts, or "brine waste" streams, are typically highly concentrated salt solutions that, in some cases, contain more than twice the amount of concentrated salts than natural brine solutions. Brine waste can also carry various contaminants, which differ depending on which process the brine waste is a byproduct of.

Brine waste streams can be some of the most challenging to treat or discharge because their composition and purification requirements can be rather dynamic and complex.

Some examples of brine waste created as a byproduct of industry include:

- Cooling tower and boiler effluent.
- Reverse osmosis (RO) and ion exchange waste/reject streams.
- Produced water from extracting oil and natural gas.
- Chlor-alkali and chemical plant waste.
- Acid rock and mine drainage.
- Food preservation and manufacturing waste streams.
- Desalination waste from potable water creation.
- Irrigation runoff.

Brine waste is typically either recycled for use in the facility's process or treated for disposal. For example, solutions with a high concentration of salt are known to reduce thermal conductivity, so brine waste is often recycled and reused as a cooling agent for steel heat exchangers in many power plants. This type of brine is often treated to remove dissolved oxygen and other harmful contaminants since brine waste can be highly corrosive to plant machinery and piping if untreated, and the presence of dissolved oxygen and other contaminants can increase that risk.

Varying salt concentrations in the brine stream will also determine how the temperature, pressure, and other threshold limitations will need to be adjusted during production, so facilities that use brine as part of their process often run frequent tests to ensure the relevant purification requirements are consistently met. It's an arduous process that often requires round-the-clock monitoring. For this reason, it's extremely important to have your water treatment specialist evaluate what the brine is being used for in addition to where and how it is being recycled and discharged to ensure its composition remains appropriate for the process or disposal at hand.

PACKAGING WASTE

Packaging waste, according to the United States Environmental Protection Agency (EPA), defined containers and packaging as products that are assumed to be discarded the same year the products they contain are purchased. The majority of the solid waste are packaging products, estimating to be about 77.9 million tons of generation in 2015 (29.7 percent of total generation). Packaging can come in all shapes and forms ranging from Amazon boxes to soda cans and are used to store, transport, contain, and protect goods to keep customer satisfaction. The type of packaging materials including glass, aluminum, steel, paper, cardboard, plastic, wood, and other miscellaneous packaging. Packaging waste is a dominant contributor in today's world and responsible for half of the waste in the globe.

The recycling rate in 2015 for containers and packaging was 53 percent. Furthermore, the process of burning of containers and packaging was 7.2 million tons (21.4 percent of total combustion with energy recovery). Following the landfills that received 29.4 million tons (21.4 percent of total land filling) within the same year.

As packaging waste pollutes the Earth, all life on Earth experiences negative impacts that affected their lifestyle. Marine or land-living animals are suffocating due to the pollution of packaging waste. This is a major issue for low income countries who do not have an efficient waste management system to clean up their environments and being the main sources for the global ocean pollution. Also, individuals who lack the motivation to recycle and instead leave their trash everywhere they went. The current location with greatest amount of solid waste that includes most of packaging products is the Great Pacific Garbage Patch located at West Coast of North America to Japan. Most packaging waste that eventually goes into the ocean often comes from places such as lakes, steams, and sewage.

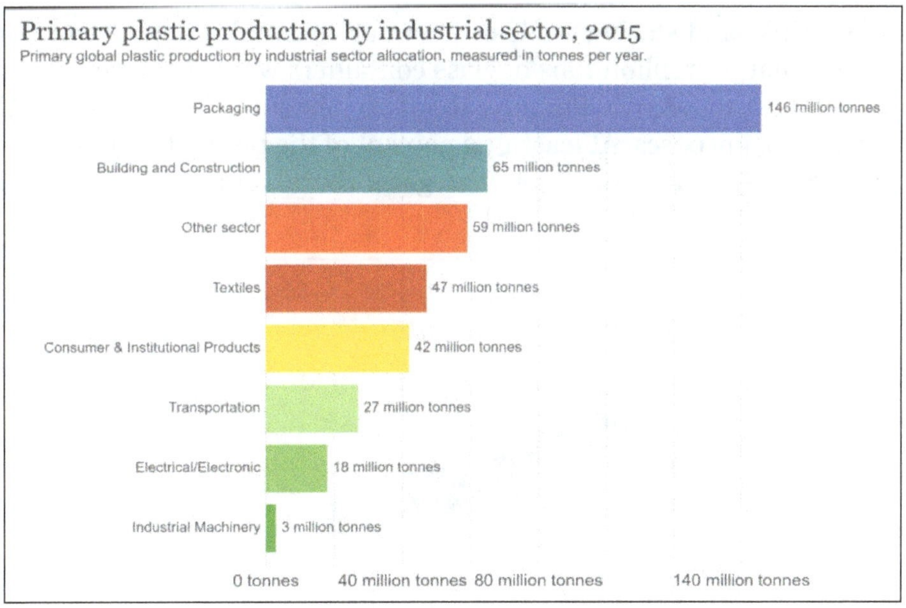

Plastic production by industrial sector.

Possible solutions to reducing packaging waste are very simple and easy and could start with minimisation of packaging material ranging up to a zero waste strategy (package-free products). The problem is mainly in a lack of motivation to start making a change. However, examples of effective ways to help reduce packaging pollution include banning the use of single-use plastics, more social awareness and education, promotion of eco-friendly alternatives, public pressure, voluntary cleaning up, and adopting reusable or biodegradable bags. Most of the solutions can be done without struggle if certain individuals are willing to do so. In reality, not many are aware or show no concern of the amount of packaging waste in our world which is very unfortunate.

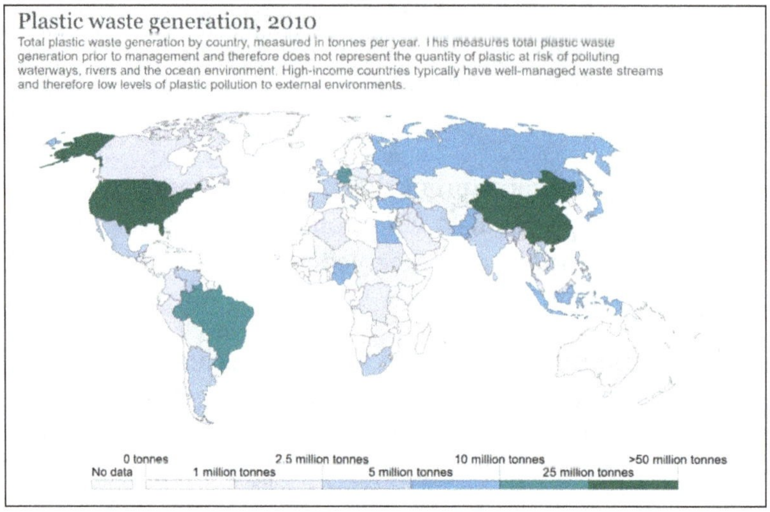

Plastic waste generation.

Types of Packaging Wastes

Glass Containers

Bottles and jars for drinks and storing foods or juices are examples of glass containers. It's been estimated by the EPA that 9.1 million tons of glass containers were generated in 2015, or 3.5 percent of municipal solid waste (MSW). Frankly, about 70 percent of glass consumption is used for containers and packaging purposes. At least 13.2 percent of the production of glass and containers are burned with energy recovery. The amount of glass containers and packaging going into the land fill is about 53 percent.

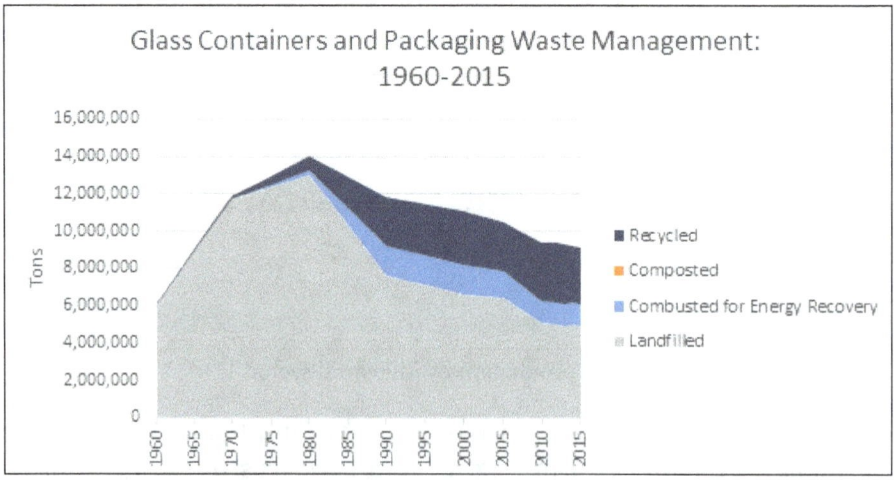

This graph represents the total number of tons of glass containers generated, recycled, composted, combusted with energy recovery and landfilled.

Aluminum Containers and Packaging

Aluminum container and packaging waste usually comes from cans from any kind of beverages, but foil can also be another that contributes it as well. It's been given that about 25 percent of aluminum is used for packaging purposes. Using the Aluminum Association Data, it has been calculated that at least 1.8 million tons of aluminum packaging were generated in 2015 or 0.8 percent MSW produced. Of those that are produced, only about 670,000 tons of aluminum containers and packaging were recycled, about 54.9 percent. And, the ones that ends up in the land fill is 50.6 percent.

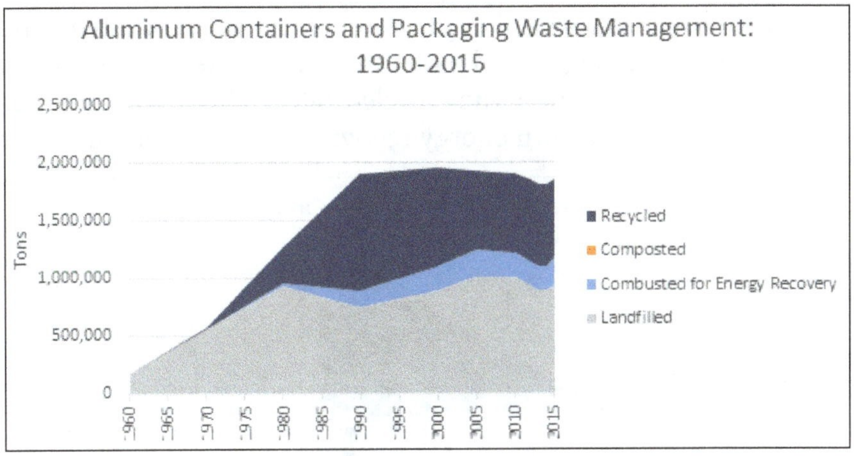

This graph represents the total number of tons of aluminum containers generated, recycled, composted, combusted with energy recovery and landfilled.

Steel Containers and Packaging

The production of steel containers and packaging mostly comes in cans and other things like steel barrels. Only about 5 percent of steel use for packaging purposes of the total world of steel consumption which makes it the least amount wasted and the most recycled. It's totaled that 2.2 million tons or 0.9 percent of MSW generated in 2015. While according to the Steel Recycling Institute, an estimate of 1.6 million tons (73 percent) of steel packaging were recycled. Adding on, the steel packaging that were combusted with energy recover was about 5.4 percent and 21.6 percent were land filled.

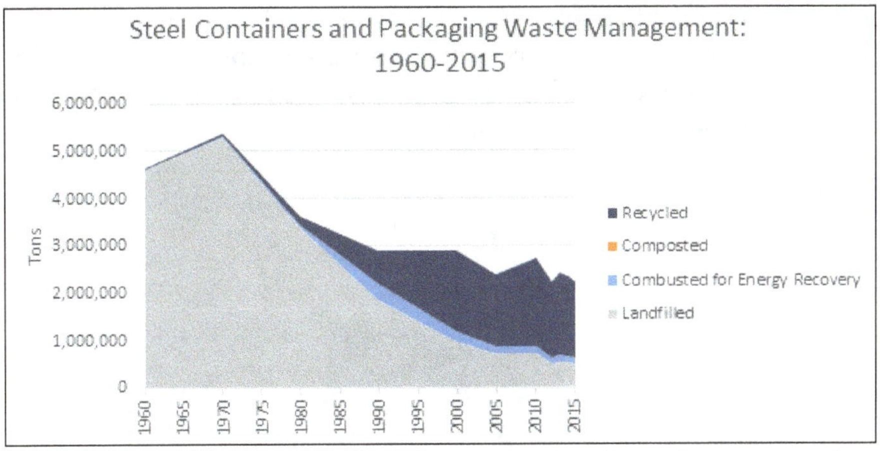

This graph represents the total number of tons of steel containers generated, recycled, composted, combusted with energy recovery and landfilled.

Paper and Paperboard Containers and Packaging

The most of it being generated, and within the MSW in 2015, was corrugated boxes coming with at least 31.3 million tons (11.3 percent total) produced. However, it also the top most recycled at 28.9 million tons (92.3 percent) boxes being recycled in 2015.

Later on, they are then combusted which makes 0.5 million tons and landfills received 1.9 million tons. Other than corrugated boxes, cartons, bags, sacks, wrapping papers, and other boxes used for shoes or cosmetics are other examples of paper and paperboard containers and packaging. The total amount of MSW generated for paper and paperboard containers and packaging was 39.9 million tons or 15.1 percent in 2015. Although, the recycled rate is about 78.2 percent and 4.3 percent of small proportions were combusted with energy recovery and 17.6 percent in landfill.

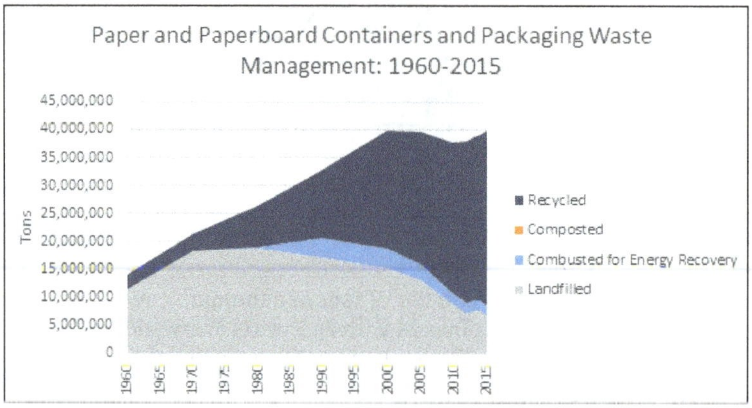

This graph represents the total number of tons of paper and paperboard containers generated, recycled, composted, combusted with energy recovery and landfilled.

Wood Packaging

Wood packaging is anything that is made out of wood used for packaging purposes (e.g. wood crates, wood chips, boards, and planks). Wood packaging is still highly use in today's world for transporting goods. According to EPA's data that were borrowed from the Virginia Polytechnic Institute and the United States Department of Agriculture's Forest Service Southern Research Station, 9.8 million tons (3.7 percent of total MSW) of wood packaging were made in production in 2015. Also, in 2015, the amount that was recycled 2.7 million tons. Moreover, its estimated that 14.3 percent of the wood containers and packaging waste generated was combusted with energy recovery, while the 58.6 percent went to the land filled.

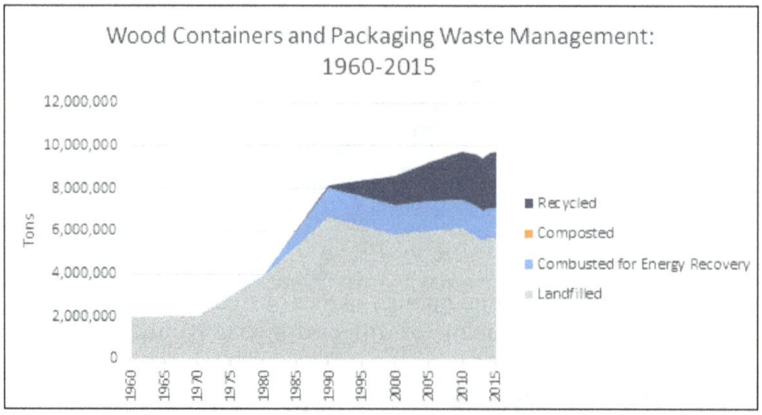

This graph represents the total number of tons of wood containers generated, recycled, composted, combusted with energy recovery and landfilled.

Plastic Containers and Packaging

Plastic containers and packaging can be found in plastic bottles, supermarket bags, milk and water jugs, and more. EPA used data from the American Chemistry Council to estimate that 14.7 million tons (5.5 percent of MSW generation) of plastic containers and packaging were created in 2015. The overall amount that is recycled is about 2.2 million tons (14.6 percent). In addition, 16.8 percent were combusted with energy recover and 68.6 percent went straight into the land fill. Most of the plastics are made from polyethylene terephthalate (PET), high-density polyethylene (HDPE), low-density polyethylene (LDPE), polyvinyl chloride (PVC), polystyrene (PS), polypropylene (PP) and other resins. That being said, the recycling rate for PET bottles and jars was 29.9 percent (890,000 tons) and the recycling of HDPE water and milk jugs was 30.3 percent (230,000 tons).

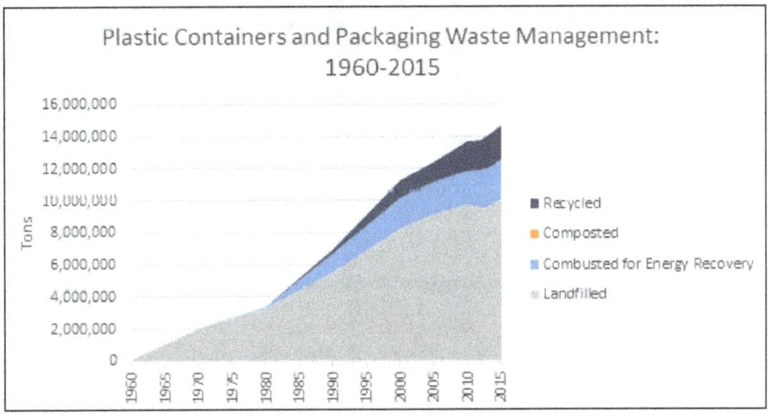

This graph represents the total number of tons of plastic containers generated, recycled, composted, combusted with energy recovery and landfilled.

Role of Packaging Waste in Pollution

Litter

Litter, a horrific sight in water systems and public places, mostly consists of packaging waste. Besides the disfigurement of the landscape, it also poses a health hazard for various life forms. Packaging materials such as glass and plastic bottles are the main constituents of litter. It has a huge impact on the marine environment as well, when animals are caught in or accidentally consume plastic packaging.

Air Pollution

The production of packaging material is the main source of the air pollution that is being spread globally. Some emissions comes from accidental fires or activities that includes incineration of packaging waste that releases vinyl chloride, CFC, and hexane. For a more direct course, emissions can originate in land fill sites which could release CO_2 and methane. Most CO_2 comes from steel and glass packaging manufacturing.

Water Pollution

Packaging waste can come from land based or marine sources. The current location that has the makes up the large of amount of water pollution is the Great Pacific Garbage Patch located at West

Coast of North America to Japan. Marine sources such as rivers that caught packaging materials and are eventually be lead to the oceans. In global standards, about 80 percent of packaging waste in ocean comes from land based sources and 20 percent comes from marine sources. The 20 percent of packaging waste that comes from marine sources comes from the rivers of China starting from least to greatest contributors, the Hanjiang, Zhujiang, Dong, Huangpu, Xi, and Yangtze river. All other marine sources comes from rivers of Africa and Southeast Asia.

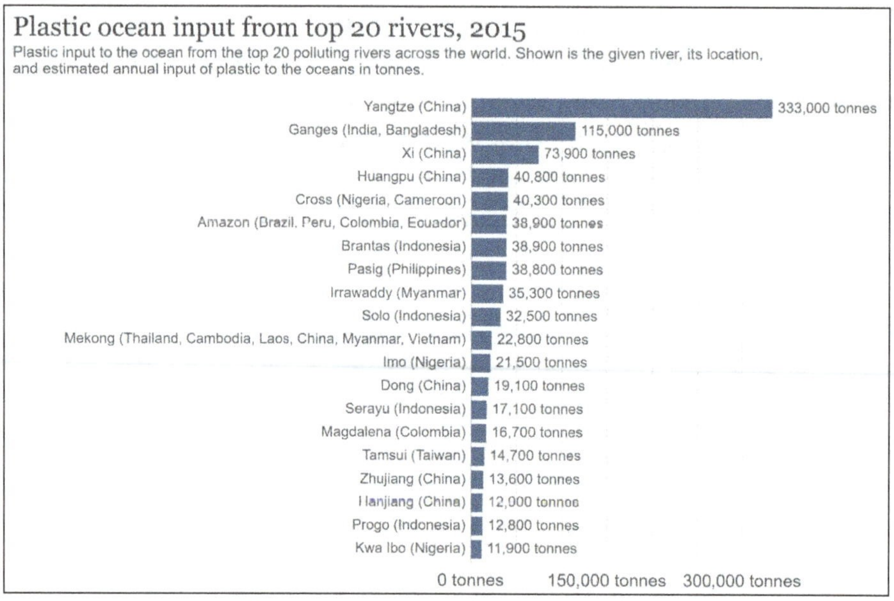

Plastic ocean input from top 20 rivers.

Impacts on Marine Species and Wildlife Species

Most marine species and wildlife species suffer from the following:

- Entanglement: At least 344 species are entangled by packaging waste, specifically the ones that are plastics. Most of the victims are marine species like whales, seabirds, turtles, and fish.
- Ingestion: 233 marine species are recorded that had consumed plastic packaging waste of either unintentionally, intentionally, or indirectly. Again, the following victims would be whales, fish, mammals, seabirds, and turtles. The effects of eating plastic packaging waste could lead to greatly reduced stomach capacity, leading to poor appetite and false sense of satiation. Whats worse is that the size of the ingested material is ultimately limited by the size of the organism. For example, microplastics consumed by planktons and fishes can consume cigarettes boxes. Plastic can also obstruct or perforate the gut, cause ulcerative lesions, or gastric rupture. This can ultimately lead to death.
- Interaction: Animals contacting with packaging waste includes collisions, obstructions, abrasions or use as substrate.

Impacts on Human Health

Bisphenol A (BPA), styrene and benzene can be found in certain packaging waste. BPA can affect the hearts of women, permanently damage the DNA of mice, and appear to be entering the human body from a variety of unknown sources. Studies from Journal of American Association shows that higher bisphenol A levels were significantly associated with heart diseases, diabetes, and abnormally high levels of certain liver enzymes. Toxins such as these are found within our food chains. When fish or plankton consume microplastics, it can also enter our food chain. Microplastics was also found in common table salt and in both tap and bottled water. Microplastics are dangerous as the toxins can affect the human body's nervous, respiratory, and reproductive system.

MINE WASTE

Mine wastes can be problematic due to the fact they contain hazardous material that can be released into the environment if not properly handled. Some of these hazardous materials include heavy metals, metalloids, radioactive waste, acidic water, and process chemicals.

Rock or Solid Mine Waste

Open pit mining has several stages of waste. First, to access the actual seams or veins of ore, the top layers of rock, or overburden, must be removed. Once the overburden has been removed, the seams can be extracted. When extracting the seams, the additional waste rock will be removed to get to the actual ore, called gangue. Then, as the material is processed and finely ground, even more, waste, called mine tailings, will need to be removed from the site.

Overburden -> Gangue -> Tailings

Overburden

Waste rock, or overburden, refers to the often large mass of initial soil and rock that is removed in order to get to the valuable mineral deposits in mining operations. Typically, for every ton of ore that is mined, 5 tons of overburden must be displaced. Overburden is not subjected to any chemical processes, but must still be removed in order to access ores. Overburden is usually managed by piling it on the nearby surface of mining sites in an area that will not interfere with ongoing operations. Transporting such large volumes of material is costly, so overburden is stored nearby and often used for revegetating the land upon closure of the mine.

Gangue

Gangue is the worthless rock or material that is closely mixed with the valuable material to be processed. The separation of mineral from gangue is called mineral processing. Often times, inefficient processing methods can produce gangue that still holds an ample amount of valuable minerals. As values of minerals increase, it can even be profitable to reprocess gangue to extract additional minerals that may have been missed during the first processing.

Mine Tailings

Tailings are finely ground rocks and other mineral waste as a result of mineral processing. Due to the way minerals are processed, tailings can contain concentrations of processing chemicals. This can make mine tailings an environmental concern, so proper transportation and disposal are crucial. Consequently, the next step is to pump mine tailings away with slurry pumps into tailings ponds. Tailings ponds are sedimentation holding ponds enclosed by dams and liners to capture and store the waste.

Liquid Mine Waste

Mine Water

Mine water is produced in a few different ways at mine sites and can vary in levels of contamination. Water exposed to mining processes is also often acidic and can contaminate local water sources in a process called acid mine drainage (AMD) or acid rock drainage (ARD). Acid mine drainage is a heavy contributor to pollution of surface water across the globe. AMD is primarily caused when water flows over the sulfide-heavy material, forming an acidic solution. Water at mine sites is usually heavily monitored and management strategies are used to not only reduce the amount of mine water produced but also to treat the water before it is released back into the environment.

Water Treatment Sludge

Water contaminated by acid rock drainage.

Sludge is produced at some mine sites and is similar to mine wastewater, but has the additions of solids and processing chemicals. These additions turn the water into a more viscous sludge which can then be pumped away from the site. Since the majority of sludge has little economic value, it's essentially handled as waste. In extreme cases where the sludge is rich in harmful or radioactive material, it may be classified as hazardous waste and require special handling and disposal methods.

References

- Industrial-waste, fact-sheets-1: safewater.org, Retrieved 19 May, 2019
- Toxic-waste, science-312021: britannica.com, Retrieved 16 July, 2019
- Introduction-chemical-waste, environmental-chemical-waste-chemical-waste-management-guide: bu.edu, Retrieved 25 June, 2019
- Mine-waste-disposal-methods, education: eddypump.com, Retrieved 28 January, 2019
- Usmani, Zeba; Kumar, Vipin (17 May 2017). "Characterization, partitioning, and potential ecological risk quantification of trace elements in coal fly ash". Environmental Science and Pollution Research. 24 (18): 15547–15566. doi:10.1007/s11356-017-9171-6
- Pongrácz, Eva (March 23, 2007), "The Environmental Impacts of Packaging", Environmentally Conscious Materials and Chemicals Processing, pp. 237–278, ISBN 9780470168219, retrieved October 30, 2019

CHAPTER 2

Sources of industrial waste include nondurable goods, durable goods, containers packaging, and food and yard waste. Other sources include battery manufacturing, organic chemicals manufacturing, food industry, iron and steel industry, etc. are the sources of industrial wastewater. The topics elaborated in this chapter will help in gaining a better perspective about these sources.

Agricultural Waste

Breweries

Beer is a fermented beverage with low alcohol content made from various types of grain. Barley predominates, but wheat, maize, and other grains can be used. The production steps include:

- Malt production and handling: Grain delivery and cleaning; steeping of the grain in water to start germination; growth of rootlets and development of enzymes (which convert starch into maltose); kilning and polishing of the malt to remove rootlets; storage of the cleaned malt.
- Wort production: Grinding the malt to grist; mixing grist with water to produce a mash in the mash tun; heating of the mash to activate enzymes; separation of grist residues in the lauter tun to leave a liquid wort; boiling of the wort with hops; separation of the wort from the trub/hot break (precipitated residues), with the liquid part of the trub being returned to the lauter tub and the spent hops going to a collection vessel; and cooling of the wort.
- Beer production: Addition of yeast to cooled wort; fermentation; separation of spent yeast by filtration, centrifugation or settling; bottling or kegging. Water consumption for breweries generally ranges 4–8 cubic meter per cubic meter (m^3/m^3) of beer produced.

Breweries can achieve an effluent discharge of 3–5 m^3/m^3 of sold beer (exclusive of cooling waters). Untreated effluents typically contain suspended solids in the range 10–60 milligrams per liter (mg/l), biochemical oxygen demand (BOD) in the range 1,000–1,500 mg/l, chemical oxygen demand (COD) in the range 1,800–3,000 mg/l, and nitrogen in the range 30–100 mg/l. Phosphorus can also be present at concentrations of the order of 10–30 mg/l. Effluents from individual process steps are variable. For example, bottle washing produces a large volume of effluent that, however, contains only a minor part of the total organics discharged from the brewery. Effluents from fermentation and filtering are high in organics and BOD but low in volume, accounting for about 3% of total wastewater volume but 97% of BOD. Effluent pH averages about 7 for the combined effluent but can fluctuate from 3 to 12 depending on the use of acid and alkaline cleaning agents. Effluent temperatures average about 30°C.

Dairy Industry

The dairy industry involves processing raw milk into products such as consumer milk, butter, cheese, yogurt, condensed milk, dried milk (milk powder), and ice cream, using processes such as chilling, pasteurization, and homogenization. Typical by-products include buttermilk, whey, and their derivatives.

Waste Characteristics

Dairy effluents contain dissolved sugars and proteins, fats, and possibly residues of additives. The key parameters are biochemical oxygen demand (BOD), with an average ranging from 0.8 to 2.5 kilograms per metric ton (kg/t) of milk in the untreated effluent; chemical oxygen demand (COD), which is normally about 1.5 times the BOD level; total suspended solids, at 100–1,000 milligrams per liter (mg/l); total dissolved solids: phosphorus (10–100 mg/l), and nitrogen (about 6% of the BOD level). Cream, butter, cheese, and whey production are major sources of BOD in wastewater. The waste load equivalents of specific milk constituents are: 1 kg of milk fat = 3 kg COD; 1 kg of lactose = 1.13 kg COD; and 1 kg protein = 1.36 kg COD. The wastewater may contain pathogens from contaminated materials or production processes. A dairy often generates odors and, in some cases, dust, which need to be controlled. Most of the solid wastes can be processed into other products and byproducts.

Pulp and Paper Industry

The pulp and paper industry is one of world's oldest and core industrial sector. The socio-economic importance of paper has its own value to the country's development as it is directly related to the industrial and economic growth of the country. Paper manufacturing is a highly capital, energy and water intensive industry. It is also a highly polluting process and requires substantial investments in pollution control equipment. The pulp and paper mill is a major industrial sector utilizing a huge amount of lignocellulosic materials and water during the manufacturing process, and releases chlorinated lignosulphonic acids, chlorinated resin acids, chlorinated phenols and chlorinated hydrocarbons in the effluent. About 500 different chlorinated organic compounds have been identified including chloroform, chlorate, resin acids, chlorinated hydrocarbons, phenols, catechols, guaiacols, furans, dioxins, syringols, vanillins, etc. These compounds are formed as a result of reaction between residual lignin from wood fibres and chlorine/chlorine compounds used for bleaching. Colored compounds and Adsorbable Organic Halogens (AOX) released from pulp and paper mills into the environment poses numerous problems. The wood pulping and production of the paper products generate a considerable amount of pollutants characterized by Biochemical Oxygen Demand (BOD), Chemical Oxygen Demand (COD), Suspended Solids (SS), toxicity, and colour when untreated or poorly treated effluents are discharged to receiving waters. The effluent is toxic to aquatic organisms and exhibits strong mutagenic effects and physiological impairment.

Iron and Steel Industry

The production of iron from its ores involves powerful reduction reactions in blast furnaces. Cooling waters are inevitably contaminated with products especially ammonia and cyanide. Production of coke from coal in coking plants also requires water cooling and the use of water in by-products separation. Contamination of waste streams includes gasification products such as benzene, naphthalene, anthracene, cyanide, ammonia, phenols, cresols together with a range of more complex organic compounds known collectively as polycyclic aromatic hydrocarbons (PAH).

The conversion of iron or steel into sheet, wire or rods requires hot and cold mechanical transformation stages frequently employing water as a lubricant and coolant. Contaminants include hydraulic oils, tallow and particulate solids. Final treatment of iron and steel products before onward sale into manufacturing includes pickling in strong mineral acid to remove rust and prepare the surface for tin or chromium plating or for other surface treatments such as galvanisation or painting. The two acids commonly used are hydrochloric acid and sulfuric acid. Wastewaters include acidic rinse waters together with waste acid. Although many plants operate acid recovery plants, (particularly those using Hydrochloric acid), where the mineral acid is boiled away from the iron salts, there remains a large volume of highly acid ferrous sulfate or ferrous chloride to be disposed of. Many steel industry wastewaters are contaminated by hydraulic oil also known as soluble oil.

Mines and Quarries

The principal waste-waters associated with mines and quarries are slurries of rock particles in water. These arise from rainfall washing exposed surfaces and haul roads and also from rock washing and grading processes. Volumes of water can be very high, especially rainfall related arisings on large sites. Some specialized separation operations, such as coal washing to separate coal from native rock using density gradients, can producewastewater contaminated by fine particulate haematite and surfactants. Oils and hydraulic oils are also common contaminants. Wastewater from metal mines and ore recovery plants are inevitably contaminated by the minerals present in the native rock formations. Following crushing and extraction of the desirable materials, undesirable materials may become contaminated in the wastewater. For metal mines, this can include unwanted metals such as zinc and other materials such as arsenic. Extraction of high value metals such as gold and silver may generate slimes containing very fine particles in where physical removal of contaminantsbecomes particularly difficult.

Food Industry

Wastewater generated from agricultural and food operations has distinctive characteristics that set it apart from common municipal wastewater managed by public or private wastewater treatment plants throughout the world: It is biodegradable and nontoxic, but that has high concentrations of biochemical oxygen demand (BOD) and suspended solids (SS). The constituents of food and agriculture wastewater are often complex to predict due to the differences in BOD and pH in effluents from vegetable, fruit, and meat products and due to the seasonal nature of food processing and postharvesting.

Processing of food from raw materials requires large volumes of high grade water. Vegetable washing generates waters with high loads of particulate matter and some dissolved organics. It may also contain surfactants.

Animal slaughter and processing produces very strong organic waste from body fluids, such as blood, and gut contents. This wastewater is frequently contaminated by significant levels of antibiotics and growth hormonesfrom the animals and by a variety of pesticides used to control external parasites. Insecticide residues in fleeces is a particular problem in treating waters generated in wool processing.

Processing food for sale produces wastes generated from cooking which are often rich in plant organic material and may also contain salt, flavourings, colouring material and acids or alkali. Very significant quantities of oil or fats may also be present.

Complex Organic Chemicals Industry

A range of industries manufacture or use complex organic chemicals. These include pesticides, pharmaceuticals, paints and dyes, petro-chemicals, detergents, plastics, paper pollution, etc. Waste waters can be contaminated by feed-stock materials, by-products, product material in soluble or particulate form, washing and cleaning agents, solvents and added value products such as plasticisers.

Nuclear Industry

The waste production from the nuclear and radio-chemicals industry is dealt with as Radioactive waste.

SOURCES OF INDUSTRIAL WASTEWATER

Battery Manufacturing

Battery manufacturers specialize in fabricating small devices for electronics and portable equipment (e.g., power tools), or larger, high-powered units for cars, trucks and other motorized vehicles. Pollutants generated at manufacturing plants includes cadmium, chromium, cobalt, copper, cyanide, iron, lead, manganese, mercury, nickel, oil and grease, silver and zinc.

Organic Chemicals Manufacturing

The specific pollutants discharged by organic chemical manufacturers vary widely from plant to plant, depending on the types of products manufactured, such as bulk organic chemicals, resins, pesticides, plastics, or synthetic fibers. Some of the organic compounds that may be discharged are benzene, chloroform, napthalene, phenols, toluene and vinyl chloride. Biochemical oxygen demand (BOD), which is a gross measurement of a range of organic pollutants, may be used to gauge the effectiveness of a biological wastewater treatment system, and is used as a regulatory parameter in some discharge permits. Metal pollutant discharges may include chromium, copper, lead, nickel and zinc.

Electric Power Plants

Fossil-fuel power stations, particularly coal-fired plants, are a major source of industrial wastewater. Many of these plants discharge wastewater with significant levels of metals such as lead, mercury, cadmium and chromium, as well as arsenic, selenium and nitrogen compounds (nitrates and nitrites). Wastewater streams include flue-gas desulfurization, fly ash, bottom ash and flue gas mercury control. Plants with air pollution controls such as wet scrubbers typically transfer the captured pollutants to the wastewater stream.

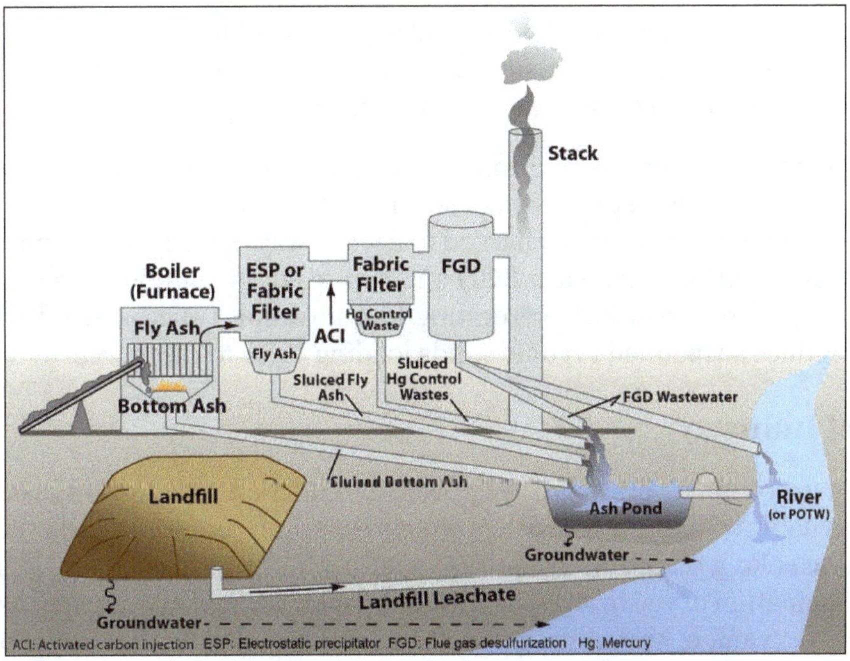

Wastestreams at a coal-fired power plant.

Ash ponds, a type of surface impoundment, are a widely used treatment technology at coal-fired plants. These ponds use gravity to settle out large particulates (measured as total suspended solids) from power plant wastewater. This technology does not treat dissolved pollutants. Power stations use additional technologies to control pollutants, depending on the particular wastestream in the plant. These include dry ash handling, closed-loop ash recycling, chemical precipitation, biological treatment (such as an activated sludge process), membrane systems, and evaporation-crystallization systems. Technological advancements in ion exchange membranes and electrodialysis systems has enabled high efficiency treatment of flue-gas desulfurization wastewater to meet recent EPA discharge limits. The treatment approach is similar for other highly scaling industrial wastewaters.

Food Industry

Wastewater generated from agricultural and food processing operations has distinctive characteristics that set it apart from common municipal wastewater managed by public or private sewage treatment plants throughout the world: It is biodegradable and non-toxic, but has high concentrations of BOD and suspended solids (SS). The constituents of food and agriculture wastewater are often complex to predict, due to the differences in BOD and pH in effluents from vegetable, fruit, and meat products and due to the seasonal nature of food processing and post-harvesting.

Processing of food from raw materials requires large volumes of high grade water. Vegetable washing generates waters with high loads of particulate matter and some dissolved organic matter. It may also contain surfactants. Dairy processing plants generate conventional pollutants (BOD, SS).

Animal slaughter and processing produces very strong organic waste from body fluids, such as blood, and gut contents. Pollutants generated include BOD, SS, coliform bacteria, oil and grease, organic nitrogen and ammonia.

Processing food for sale produces wastes generated from cooking which are often rich in plant organic material and may also contain salt, flavourings, colouring material and acids or alkali. Very significant quantities of oil or fats may also be present.

Food processing activities such as plant cleaning, material conveying, bottling, and product washing create wastewater. Many food processing facilities require on-site treatment before operational wastewater can be land applied or discharged to a waterway or a sewer system. High suspended solids levels of organic particles increase BOD and can result in significant sewer surcharge fees. Sedimentation, wedgewire screening, or rotating belt filtration (microscreening) are commonly used methods to reduce suspended organic solids loading prior to discharge.

Iron and Steel Industry

The production of iron from its ores involves powerful reduction reactions in blast furnaces. Cooling waters are inevitably contaminated with products especially ammonia and cyanide. Production of coke from coal in coking plants also requires water cooling and the use of water in by-products separation. Contamination of waste streams includes gasification products such as benzene, naphthalene, anthracene, cyanide, ammonia, phenols, cresols together with a range of more complex organic compounds known collectively as polycyclic aromatic hydrocarbons (PAH).

The conversion of iron or steel into sheet, wire or rods requires hot and cold mechanical transformation stages frequently employing water as a lubricant and coolant. Contaminants include hydraulic oils, tallow and particulate solids. Final treatment of iron and steel products before onward sale into manufacturing includes *pickling* in strong mineral acid to remove rust and prepare the surface for tin or chromium plating or for other surface treatments such as galvanisation or painting. The two acids commonly used are hydrochloric acid and sulfuric acid. Wastewaters include acidic rinse waters together with waste acid. Although many plants operate acid recovery plants (particularly those using hydrochloric acid), where the mineral acid is boiled away from the iron salts, there remains a large volume of highly acid ferrous sulfate or ferrous chloride to be disposed of. Many steel industry wastewaters are contaminated by hydraulic oil, also known as soluble oil.

Mines and Quarries

Mine wastewater effluent in Peru, with neutralized pH from tailing runoff.

The principal waste-waters associated with mines and quarries are slurries of rock particles in water. These arise from rainfall washing exposed surfaces and haul roads and also from rock washing and grading processes. Volumes of water can be very high, especially rainfall related arisings on large sites. Some specialized separation operations, such as coal washing to separate coal from native rock using density gradients, can produce wastewater contaminated by fine particulate haematite and surfactants. Oils and hydraulic oils are also common contaminants.

Wastewater from metal mines and ore recovery plants are inevitably contaminated by the minerals present in the native rock formations. Following crushing and extraction of the desirable materials, undesirable materials may enter the wastewater stream. For metal mines, this can include unwanted metals such as zinc and other materials such as arsenic. Extraction of high value metals such as gold and silver may generate slimes containing very fine particles in where physical removal of contaminants becomes particularly difficult.

Additionally, the geologic formations that harbour economically valuable metals such as copper and gold very often consist of sulphide-type ores. The processing entails grinding the rock into fine particles and then extracting the desired metal(s), with the leftover rock being known as tailings. These tailings contain a combination of not only undesirable leftover metals, but also sulphide components which eventually form sulphuric acid upon the exposure to air and water that inevitably occurs when the tailings are disposed of in large impoundments. The resulting acid mine drainage, which is often rich in heavy metals (because acids dissolve metals), is one of the many environmental impacts of mining.

Nuclear Industry

The waste production from the nuclear and radio-chemicals industry is dealt with as Radioactive waste.

Petroleum Refining and Petrochemicals

Pollutants discharged at petroleum refineries and petrochemical plants include conventional pollutants (BOD, oil and grease, suspended solids), ammonia, chromium, phenols and sulfides.

Pulp and Paper Industry

Effluent from the pulp and paper industry is generally high in suspended solids and BOD. Plants that bleach wood pulp for paper making may generate chloroform, dioxins (including 2,3,7,8-TCDD), furans, phenols and chemical oxygen demand (COD). Stand-alone paper mills using imported pulp may only require simple primary treatment, such as sedimentation or dissolved air flotation. Increased BOD or COD loadings, as well as organic pollutants, may require biological treatment such as activated sludge or upflow anaerobic sludge blanket reactors. For mills with high inorganic loadings like salt, tertiary treatments may be required, either general membrane treatments like ultrafiltration or reverse osmosis or treatments to remove specific contaminants, such as nutrients.

Textile Mills

Textile mills, including carpet manufacturers, generate wastewater from a wide variety of processes, including wool cleaning and finishing, yarn manufacturing and fabric finishing (such as bleaching, dyeing, resin treatment, waterproofing and flameproofing). Pollutants generated by textile mills include BOD, SS, oil and grease, sulfide, phenols and chromium. Insecticide residues in fleeces are a particular problem in treating waters generated in wool processing. Animal fats may be present in the wastewater, which if not contaminated, can be recovered for the production of tallow or further rendering.

Textile dyeing plants generate wastewater that contain synthetic and natural dyestuff, gum thickener (guar) and various wetting agents, pH buffers and dye retardants or accelerators. Following treatment with polymer-based flocculants and settling agents, typical monitoring parameters include BOD, COD, color (ADMI), sulfide, oil and grease, phenol, TSS and heavy metals (chromium, zinc, lead, copper).

Industrial Oil Contamination

Industrial applications where oil enters the wastewater stream may include vehicle wash bays, workshops, fuel storage depots, transport hubs and power generation. Often the wastewater is discharged into local sewer or trade waste systems and must meet local environmental specifications. Typical contaminants can include solvents, detergents, grit, lubricants and hydrocarbons.

Water Treatment

Many industries have a need to treat water to obtain very high quality water for demanding purposes such pure chemical synthesis or boiler feed water. Many water treatment produce organic and mineral sludges from filtration and sedimentation. Ion exchange using natural or synthetic resins removes calcium, magnesium and carbonate ions from water, typically replacing them with sodium, chloride, hydroxyl and other ions. Regeneration of ion exchange columns with strong acids and alkalis produces a wastewater rich in hardness ions which are readily precipitated out, especially when in admixture with other wastewater constituents.

Wood Preserving

Wood preserving plants generate conventional and toxic pollutants, including arsenic, COD, copper, chromium, abnormally high or low pH, phenols, oil and grease, and suspended solids.

CHAPTER 3

Industrial waste impacts the environment adversely as it can be toxic, ignitable, reactive and radioactive. Such waste pollutes the air, soil and nearby water sources thereby affecting the marine wildlife and the ecosystem. This chapter delves into the impacts of industrial waste to provide an in-depth understanding of the subject.

HARMFUL EFFLUENTS INDUSTRIAL DISCHARGE

Acids

Various acid manufacture industries discharge acidic effluent which not only make the land infertile but as make the water of the river acidic. The high acidity has caused stomach diseases and skin ailments in the bodies of human beings.

Discharge of Nitrogen Dioxide

Many industries release SO_2 in the air as pollutant which on reaction with rain water converts into sulphurous acid and gets precipitated in the fields and rivers thus destroying the vegetation and drinking water sources.

These are some industries which discharge nitrogen dioxide also in the air and so cause precipitation of nitrous acid in the fields and rivers which ultimately cause pollution. This pollution has also caused respiratory disorders and other ailments of lungs and heart in the human body.

Alkalies

There are alkali manufacture or textile industries which discharge alkaline effluents causing infertility of the soil and render the water useless for human beings. They destroy the flora and fauna of the vicinity.

Pesticides

There are thousands of pesticides in the market. They either contaminate the water by phosphate or chloride along with increase of BOD, COD, Sulphate and nitrates. There are some pesticides which cause BOD even to the range of 20,000 – 30,000 mg/1. The dangerous pesticides such as DDT, aldrin, dieldrin, heptachlor, benzene etc. are harmful for the aquatic life and human beings.

Some years back due to leakage of methyl isocyanate from a pesticide factory at Bhopal, about 20,000 persons were crippled and about 15,000 died within a couple of days.

Toxic Chemicals

The chemical industries discharge various toxic chemicals in the fields causing destruction of vegetation, soil and water. They should discharge effluent after proper treatment otherwise the life on this planet will be difficult. They release even cyanides in the fields due to which at one time about 50,000 fishes were died in in 1984.

Fluoride

These industries discharges lot of fluorides in the fields without any treatment due to which people suffer from mental and stomach ailments. Fluoride also reduces production in fields.

Toxic Metals

The metal industries discharge various metals, metallic oxides, other slag causing metallic pollution in the air, water and soil. The toxic metals percolate in the ground water slowly and slowly and cause enteric diseases and mental disorders.

Fertilizers

The fertilizer industries either discharge sulphur dioxide or nitrogen dioxide in the air and both cause acid rain or destroy the growth of trees, fruits, vegetables and even grasses. The effluent of these industries is also harmful and must be discharged after proper treatment.

Oil

Oil industry discharge oil directly into the sea due to which sea kingdom has been completely disturbed. Marine pollution must be stopped immediately. In short, all big industries such as Rubber, Textile, Rayon, Steel, Titanium dioxide etc. release harmful pollutants in the air and discharge toxic effluents which are harmful to aquatic life and human beings.

INDUSTRIAL POLLUTION

Industrial pollution refers explicitly to any contamination caused by industrial activities. Industrial pollution is a big issue because most pollution is caused by some industry, making it the most significant form of pollution on the planet.

The effects of industrial pollution are vast, causing water contamination, a release of toxins into soil and the air, and it is the cause of some of the most significant environmental disasters of all time.

Six Causes of Industrial Pollution

The causes of industrial pollution are extensive but here is a list of some of the top reasons.

Industrial Growth that is Unplanned

A lot of air and water pollution has occurred from companies who ignored rules or standard practices to facilitate rapid growth. Industrial growth has been a frequent culprit for pollution.

Lack of Effective Policies

Many industries have been able to ignore or entirely bypass pollution laws because the policies are either not valid, or not adequately enforced by pollution control boards. The result has been significant pollution that has had a direct effect on many people.

Sheer Number of Industries

There are many industries and factories, and most of them are polluting the air every day. Small sectors that have to rely on grants from the government to keep operating are often able to avoid following environmental regulations. Those companies have been discovered to release significant amounts of toxic gas, making pollution an even more substantial issue from those smaller industries.

Using Old and Outdated Technology

Updating technology is expensive, and there are many companies and plants which are relying on outdated technology to continue operating their business successfully. Older technologies tend to produce large amounts of waste.

Even when companies realize the harmful effect they are having on the environment, they choose to save money by keeping their old systems instead of upgrading.

Natural Resource Use

Raw material is necessary for a lot of industries, which requires them to pull underground elements. One of the most common forms of leaching from natural resources is fracking for oil. When industries pull minerals, the process causes pollution in the soil and also causes oil leaks and spills that are harmful and deadly to people and animals.

Improper Disposal of Waste

One of the most common forms of soil and water pollution are because of companies who do not dispose of their waste correctly. This is one of the most significant causes of pollution because the effects include severe and chronic health issues and lower air quality.

The causes of industrial pollution are extensive but here is a list of some of the top reasons.

Effects of Industrial Pollution

Water Pollution

The ecosystem has been significantly impacted because of pollution, and the consequence will affect many future generations. Big industries and manufacturing plants use significant amounts of water from nearby lakes, rivers, and oceans to operate.

During the manufacturing process, the water is exposed to chemicals, heavy metals, organic sludge, and radioactive waste. That water is then dumped back into its original water source. Not only does this process affect the plants, animals, and insects in their habitat, farmers also use that water for irrigation, causing harmful effects to the food we eat.

Water pollution has caused many different groundwater sources to be labeled as unsafe for people and animals alike. Once it is rendered dangerous, the only use for that water is for it to be reused in the plants that initially harmed it.

Soil Pollution

Soil pollution is the result of land degradation, and it is caused because of human-made (xenobiotic) chemicals as a result of industrial activity, the improper disposing of waste and various agricultural chemicals. Soil pollution poisons agriculture and sometimes kills local vegetation.

The results of soil pollution include chronic health concerns for people who work with soil, for farmers, and for people who consume the fruits and vegetables from contaminated land.

Air Pollution

Similar to water and soil pollution, air pollution has been the culprit for many illnesses, increasingly frequent over time. The effects of air pollution have an impact on everyone, every day. It is caused by the smoke released by various industries.

Animal Extinction

Pollution is one of the most significant contributors to global warming. There are conspiracy theories and people who believe global warming doesn't exist, but scientists and environmentalists have proven the existence of global warming.

The rapid increase of global warming is of concern to many people. The greenhouse gasses, and the smoke that causes it is released into the air daily and causing consequential problems. Glaciers are melting, polar animals are becoming extinct, tsunamis, hurricanes, floods, and other natural disasters are increasing, and all of those are the effects of global warming.

Natural Resource Use

Industrial pollution continues to cause significant damage to the earth and all of its inhabitants. It disrupts natural habitats and rhythms, affecting wildlife and ecosystems. Animals are becoming extinct, and habitats are being destroyed.

Pollution is the culprit for oil spills and radioactive material leaks, and both of those types of disasters take years to decades to clean up.

Industrial pollution is wreaking havoc on the planet. Every nation is affected, and there are many people who are working tirelessly to increase awareness and advocate for change. The activities that are causing pollution include:

- Burning fossil fuels like oil, natural gas, and petroleum.
- Burning coal.
- Chemical solvents used in dyeing and tanning industries.
- Untreated gas and liquid waste being released into the environment.
- Improper disposal of radioactive material.

Reducing Industrial Pollution

As the harmful effects of industrial pollution increase, there are many agencies and individuals who are working to reduce carbon footprints and live and work eco-friendlier. Here are some of the different steps being taken to reduce and eliminate pollution:

- Development of better technology for waste disposal.
- Increased recycling efforts.
- Development of cooling rooms or bins that allow industries to recycle the water they need instead of pushing it back into the natural water source it came from.
- Adopting organic water and soil cleaning methods, like using microbes that feed off of metal and waste.
- Creating policies that prevent land misuse.

Other Ways to Control or Reduce Industrial Pollution

Here are some of the ways companies are working to reduce industrial pollution.

Source Control

Adopting new technology, efficiently training employees for safe use and disposal of waste, and being more conscientious about the raw materials used can help control industrial pollution at the source.

Industry Site Selection

Putting more consideration into where new sites are located, and the potential impact of the environment surrounding it can help reduce harmful consequences.

Proper Treatment of Industrial Waste

By developing and implementing adequate treatment habits for handling industrial waste, pollution can be reduced.

Planting and Rebuilding

Rebuilding habitats by planting more trees and plants can help give wildlife back their homes and the trees can help purify air, output oxygen, and act as a buffer against the environment.

Stricter Laws and Enforcement

While we do have the Environmental Protection Agency (EPA) working to correct the damage from pollution, they need more ability to take action against the companies who do not follow proper protocol. Rule enforcement needs to have stronger consequences and more significant rewards for the companies who do operate properly.

Regular Environmental Impact Assessments

Being a responsible company or industry should require regular environmental impact assessments that are reported for evaluation. If there are harmful impacts discovered during the review, actions to correct the negative consequences should be developed and enforced.

Soil Pollution by Agricultural and Industrial Waste

Soil contaminated by heavy metals from agricultural and industrial wastes will produce unhealthy food. Heavy metals enter the food chain and are consumed by human beings.

In modern economies, various types of activity, including agriculture, industry and transportation, produce a large amount of wastes and new types of pollutants. Soil, air and water have traditionally been used as sites for the disposal of all these wastes. For example, beef cattle in the United States are estimated to produce 92 million mt/year of manure, while dairy cattle produce 27 million mt/year. Some of this manure may wash into nearby streams, and pollute rivers, lakes and soil.

The most common kinds of waste can be classified into four types: agricultural, industrial, municipal and nuclear. Agricultural wastes include a wide range of organic materials (often containing pesticides), animal wastes, and timber by-products. Many of these, such as plant residues and livestock manure, are very beneficial if they are returned to the soil. However, improper handling and disposal may cause pollution.

Industrial waste products may be in gas, liquid or solid form. The most important gases are carbon dioxide (CO_2), carbon monoxide (CO), nitrogen dioxide (NO_2) and sulfur dioxide (SO_2). They are produced by combustion in industry and by automobiles, and they pose a hazard to the environment.

Food processing plants produce both liquid and solid wastes. Another urban waste is municipal garbage. This is made up of materials discarded by homes and industry. It contains paper, plastic and organic materials. Some of these can be recycled by composting or they may be burnt or disposed of in landfills.

Sewage sludge is the product of treatment plants. The materials processed in the treatment plants are domestic and industrial wastes. They are usually liquid mixtures, composed both of solids, and of dissolved organic and inorganic materials. The water is separated from the solid part by a number of treatments before it is environmentally safe for discharge into streams or lakes.

The content of major nutrients and micronutrients in sewage sludge varies depending on the source. Data indicates that the nitrogen content of textile sludge is generally high. However, the heavy metal content is also high. Some trace elements are required in small amounts by plants and animals, whereas others are hazardous to human health.

Contamination from Industrial Wastes

Areas are being polluted by heavy metals from sewage sludge produced by the textile industry. This waste is disposed of directly into three rivers, all of which are used to irrigate lowland rice. About 720 ha of lowland rice fields were polluted in this way.

Soil surveys revealed that there were very high concentrations of boron, cadmium and lead in three villages in the Rancaekek district. Falling soil productivity in these areas caused a reduction in rice yields and farmers' incomes. After 20 years of contamination, the average rice yield had decreased by about 80%. The initial rice yield of about 4-6 mt/ha had become 1 mt/ha. However, the heavy metal content in the soil had increased by about 18% - 98%, compared to unpolluted soil.

A greenhouse study using polluted soil from this area showed that high concentrations of lead, cadmium, copper, chromium and boron were found in the plant tissue, roots and grain of rice. Most of the pollutants had accumulated in the root system.

Contamination from Agricultural Wastes

A study found that the levels of lead and cadmium in the soil were fairly low. Lead was present in soil samples in a range of 10 - 43 ppm, while the levels of cadmium were 0.19 - 0.49 ppm.

The content of lead and cadmium which were present may have originated in applications of phosphate fertilizer. The cadmium content of phosphate fertilizer in Indonesia is 35 - 255 g/mt.

Phosphate fertilizer is essential in intensive agriculture, especially in Indonesia with its high rainfall and rapid leaching. These conditions result in a low soil pH and high levels of iron and aluminum oxide. These in turn immobilize the phosphorus in the soil solution, and hinder its uptake by plants.

Table: Total area of intensive paddy rice contaminated by lead and cadmium in Kawarang and Bekasi District, West Java, Indonesia.

Distance from main road (m)	Lead		Cadmium	
	Plantation A	Plantation B	Plantation A	Plantation B
0	37	55	0.6	0.4
50	24	30	-,8	0.4
100	23	28	0.7	0.4
200	19	27	0.6	0.4

Based on the levels of lead and cadmium in rice, Kasno found that intensive lowland rice areas could be divided into three categories: Highly polluted soils, soils with medium pollution, and unpolluted soils. Only 7% of the total lowland areas studied were polluted by lead, and about 4% by cadmium. These results indicate that after 30 - 40 years of phosphate application, the productivity of these soils could still be sustained.

Table: Heavy metal content in two tea plantations, West Java, Indonesia.

Villages	Distance from traditional mining loction (km)	Mercury contant (ppm)		
		Soil	Rice straw	Rice grain
A, B	< 0.1	6.7	5.3	0.43
C	0.8 – 1.0	5.6	1.8	< 0.0005
D	1.2 – 1.5	1.8	0.8	< 0.0005
E	7.0 – 7.5	2.4	-	< 0.25
F	11.5 – 12.0	1.3	-	< 0.0005

Another study was conducted in tea plantations which is important for agroforestry and tourism. The aim of the study was to see the effect of air pollution by automobiles on soil quality. The result of the soil survey showed that the lead content of the soil in the plantations increased near main roads. The level of soil pollution by lead, most of which was produced by petrol combustion, depended on the distance from the main road. However, the cadmium content in soils was not influenced by the distance from the main road. This indicates that the cadmium content in the soil was not the result of air pollution, but may have resulted from the application of high levels of phosphate fertilizer in these areas.

Contamination from Gold Mining and Smelting

Gold mining is carried out by individuals rather than companies. They use traditional methods for separating the gold from the raw material. The main waste product from this process is mud and rubble which contain a high concentration of mercury. These wastes are disposed of directly in the river, which is also used as a source of irrigation water in the lowland rice areas around the mining areas.

Table: Mercury content of soil, rice straw and rice grain in area contaminated by gold mining

Treatment		Site A				Site B	
	Before		After		Before		After
				Lead (ppm)			
Vetiver	38		15		14		8
Vetiver + 80 ppm lead	118		48		94		58
				Lead (ppm)			
Vetiver	1.2		0.7		0.6		0.4
Vetiver + 20 ppm cadmium	21.1		12.7		20.6		13.4

A soil survey found that the soil surrounding the traditional mining was polluted by mercury. The pollution covered the land around six villages. The concentration of mercury in soil near the mining was higher than in more distant soils. A high concentration of mercury was found in rice straw and rice grain. All of the values were higher than the maximum permitted level of mercury in soils (0.5 ppm).

IMPACT OF INDUSTRIAL WASTES ON AQUATIC LIFE

Impacts on Marine Life

Marine pollution occurs when harmful effects, or potentially harmful effects, can result from the entry into the ocean of chemicals, particles, industrial and residential waste, noise or the spread of invasine organisms. Most sources of Marine pollution are land based.

Many potentially toxic chemicals adhere to tiny particles which are then taken up by plankton and benthos animals, most of which are either deposit or filter feeders. In this way, the toxins are concentrated upward with in ocean food chains. Many particles combine chemically in a manner highly depletive oxygen, causing estuaries to become anoxic.

When pesticides are incorporated into the marine ecosystem, they quickly become absorbed into marine food webs. Once in the food webs, these pesticides can cause mutations, as well as diseases, which can be harmful to humans as well as the entire food web. These toxic metals can also be introduced into marine food webs.

These can cause a change to tissue matter, biochemistry, behaviour, reproduction and suppress growth in marine life. Also, many animal feeds have a high fish meal or fish hydro lysate content. In this way, marine toxins can be transferred to land animals, and appear later in meat and dairy products.

There are different paths of pollution:

- Direct Discharge.

- Land Runoff.
- Ship Pollution.
- Atmospheric Pollution.
- Deep Sea Mining.

Important Concepts Related to Marine Pollution

Ocean Acidification

The oceans are normally a natural carbon sink, absorbing carbon dioxide from the atmosphere. Since the levels of atmospheric CO_2 are increasing, the oceans are becoming more acidic. The potential consequences of ocean acidification are not fully understood, but there are concerns that structures made of calcium carbonate may become vulnerable to dissolution, affecting corals and ability of shellfish to form shells.

Eutrophication

Eutrophication is an increase in chemicals nutrients, typically compounds containing nitrogen or phosphorus, in an ecosystem. It can result in an increase in the ecosystems primary productivity (excessive plant growth and decay), and further effects including lack of oxygen and severe reductions in water quality, fish and other animal populations. An excess of oxygen depleting chemicals in the water can lead to hypoxia and creation of a dead zone.

Impacts on Fresh Water

Fresh Water makes up less than three percent of earths water, but is the source of virtually all drinking water. According to the survey published by Environmental Protection Agency (EPA) in 2000, 40 percent of U.S. rivers and 45 percent of lakes are polluted. These sources, called surface water, are vulnerable to pollution discharged out of pipes and precipitating out of the air but the primary source of their pollution today is runoff, pollutants washing off the land.

Industrial waste water (mixture of toxic chemicals) and domestic sewage caused depletion of oxygen from the water column as it decomposes, stress or suffocating aquatic life. Heavy metals from industries and mining sites (like, Pb, Cd, Zn, CO, Ni, Hg, As etc.) persist in fresh water environments such as river sediments and wetlands for long periods are accumulated in the tissues of fish and shellfish. These are highly toxic to both aquatic organisms and humans who consume them.

The concentration of dissolved O_2 highly depends on the amount of pollutants, because most water pollutants cause low oxygen levels in fresh water. These pollutants make it difficult for species to live, any many, aquatic organisms, especially fish, die when dissolved oxygen levels fall below 5 ppm.

Mostly, freshwater ecosystems may become unbalanced by factors due to human activities and industrial wastes. All these affect the bioavailability of chemicals to organisms, cause temperature fluctuations, and modify rainfall, pH and salinity. Some 1500 substances have been listed as pollutants in fresh water ecosystems.

Sources of water pollution can be grouped into two categories based on their origin:

Point Sources

Point source water pollution refers to contaminants that enter a waterway from a single, indentifiable source, such as a pipe or ditch. Examples of sources in this category include discharges from a sewage treatment plant, a factory, a city storm drain.

Non-point Sources

Non-point source pollution refers to diffuse contamination that does not originate from a single discrete source. NPS pollution is after the communicative effect of small amounts of contaminants gathered from a large area. A common example is the leaching out of nitrogen compounds from fertilized agricultural lands. Nutrient run off in Strom water from "sheet flow" cover an agricultural field as a forest are also cited as examples of NPS pollution.

EFFECTS OF INDUSTRIAL EFFLUENT ON SOIL CHARACTERISTICS

An effluent is an inevitable production of industrial process. It is defined by the United States Environmental Protection Agency as "wastewater (treated or untreated) that flows out of a treatment plant, sewer or industrial outfall. Increased number of industries has enlarged the disposal of effluent to open land or to natural water resources. Effluent of different industries may vary in composition depending upon the source of production. Effluent may contain essential nutrients and some toxic substances. The available macronutrients and micronutrients of effluents can increase soil fertility. On the other hand the heavy metals and toxic components can accumulate in soil. The various effects depend on the period of effluent application. Metal contaminated soils can cause health problems to animals and human beings when such plants are consumed. Many investigators stated that heavy metals get accumulated in soil resulted from continuous irrigation with sewage effluent. On the other hand the use of waste water could solve the disposal problems, minimizes the risk of public health and also prevents the ground water pollution. Wastewaters from urban and agricultural sources have great potential for reuse as sources of water, organic matter, nutrients and soil conditioning agents.

On an average, the soil pH dropped by 0.3 U as a result of sewage irrigation. Effluent irrigation reduced the soil bulk density and soil pH, but had no effect on soil infiltration rates. It increased the soil nutrient concentrations and organic matter content in many of the researches. The effluent irrigation increased the total biomass production of the plantation. Soil properties are hardly affected by different plant species. The species variation like planting Eucalyptus botryoides, E. globulus, and E. ovata had limited influence on soil change. Authors also found decrease in bulk density along with increase in organic matter and nutrients in soil. They concluded the increase in total biomass production of the plantation by using effluent water. Use of sewage for irrigation in various proportions improved the organic matter to 1.24–1.78% and fertility status of soils especially down to a distance of 1 km along the disposal channel. Build up in total N was up to 2908 kg/ ha, available P (58 kg/ ha), total P (2115 kg/ ha), available K (305 kg/ ha) and total K (4712 kg/ ha) in surface 0.15 m soil. Mixing of effluent from different sources can give advantageous results. Addition of municipal effluent to textile effluent is recommended as one of the better options for utilization of textile industrial effluent in tree growing. Seedlings irrigated with municipal effluent showed optimum growth and plant nutrient concentration. In the long term, however, the changes in soil exchangeable cations may affect soil quality and the sustainability of plant production, and could be a problem for sustainability of effluent application. The N-reduced effluent had a more profound effect on the concentrations of cations in the drainage water than the primary-treated effluent. It is concluded that application of meat processing effluents can increase plant production due to increased nutrient loadings. The effects of the non-N wastewater components (e.g. cations, particularly sodium) and hydraulic loading rates on soil properties and waterways are also critical for the sustainability of wastewater irrigation. Sodium (Na^+) is of particular concern as it accumulates in soil and displaces other cations such as calcium (Ca^{2+}) and magnesium (Mg^{2+}). Excessive Na^+ may also have an adverse effect on soil structure, causing a decline in soil permeability, and therefore in a soil's ability to further accept and treat wastewater.

The waste waters carry appreciable amounts of trace toxic metals and concentrations of trace metals in sewage effluents vary from city to city. Concentration of heavy metals in sewage effluents are low, long-term use of these waste waters on agricultural lands often results in the build-up of the elevated levels of these metals in soils. Significant positive correlations were found between the concentrations of nutrients in soil and tree leaf suggest that the trend of mineral concentration in plants would be a function of mineral content in soil and also depended on their uptake by plants. In the absence of heavy metals, vegetables grown on sewage irrigated soils are still safe to be consumed by human being. Crop species exercise differently in accumulating heavy metals of effluents irrigated soils in their tissue. Uptake of heavy metals by plants may be an indicator of efficiency of metal absorption of various crop species grown on metal contaminated soils. Sewage irrigation for 20 years resulted into significant build-up of DTPA extractable Zn (208%), Cu (170%), Fe (170%), Ni (63%) and Pb (29%) in sewage-irrigated soils in vicinity of Delhi over adjacent tube well water irrigated soils, whereas Mn was depleted by 31%.

Soils receiving sewage irrigation for 10 years exhibited significant increase in Zn, Fe, Ni and Pb, while only Fe in soils was positively affected by sewage irrigation for 5 years. Rai (Brassica juncea) and tori (Brassica campestris) accumulated higher content of heavy metals especially Ni as compared to other crops, with higher content in roots than shoots. There are some reports which indicate that heavy metals pollution including Ni in soils arises as a result of various anthropogenic activities such as continuous use of sewage water and sewage sludge. The content of DTPA –Ni showed decreasing trend with depth. Ni is translocated to top parts of plants in the given order tori > rai> maize> bajra >lady finger. Trace elements including Ni are present in different chemical forms because they are associated to diverse organic and inorganic components of the soils such as soluble, exchangeable, bound to carbonate, Fe and Mn oxides, organic matter and residual. With the exception of Zn in plant tissues harvested from compost and sludge treated soils, root tissue of cabbage for both compost and sludge treated treatments appeared to accumulate more metals in the roots than the aerial parts.

The effect of irrigation with secondary treated municipal effluent on the accumulation of heavy metals (Cd, Cr, Cu, Ni, Pb, and Zn) was investigated by monitoring sites that had been irrigated with effluent for 4 and 17 years. Concerning nutrient recycling, it is commonly admitted that wastewater irrigation results in higher yield than irrigation with fresh water due to the supply of nutrients (especially nitrogen, Phosphorus and potassium) contained in the wastewaters. Chemical analysis clearly exhibited sodium and bicarbonates accumulation, and pH increase to alkaline values, in the topsoil of the plots irrigated with recycled wastewater. Shrinkage analysis showed that the structural pore network of the soil collapsed dramatically, resulting in compact layers with poor water storage. In addition, alkalinization induced organic matter dissolution resulting in black alkali forming at the soil surface. The level of sanitary hazard involved in agricultural use of wastewater depends on the nature of the effluent treatment prior to reuse. There are different views associated with effects of long-term waste water application on soil pH. There are studies which confirms increase in soil pH with waste water irrigation of soil.

Similar results showed of decrease in soil pH with waste water irrigation in different soils of varying agro-climatic conditions have also been reported. Elements such as Ca and Mg needed for plant growth can accumulate in soils thereby improving the pH especially of acidic soils. It can be concluded that some industrial effluents hold great promise for the improvement of soil fertility and crop yields, if proper treatment and management practices are adopted. Safe utilization of waste water for irrigation to crops requires several precautionary measures viz. adequate dilution, selection of crop etc. Besides, soil physical properties needs to be reviewed periodically for long-term sustainability of the system. Long-term use of waste water for irrigation to crops results in significant increased in soil organic content than the soils irrigated with ground water.

The concentrations of ions in soils are influenced by water movement, their concentrations in irrigation water, soil properties and plant uptake. Phosphorous is an immobile nutrient in soil, thus long term field studies are necessary to understand the effluent-P dynamics in the soil. Land application of wastewaters can increase the levels of soluble and exchangeable forms of potassium in soil more rapidly than with conventional inorganic fertilizers. Most of the potassium in wastewater is immediately available. The quantity of heavy metal absorption using plant depends upon many factors including the total quantity of the elements applied through sewage application, soil properties, and type of plant. It was found overall concentrations of all measured elements except Calcium and Cadmium were significantly elevated in the soil that was irrigated with effluent. Sodium concentration and pH were elevated in the upper soil horizon, whereas Phosphorus concentrations were elevated uniformly.

Differences in electrical conductivity and Magnesium concentrations between the effluent-irrigated field and the unirrigated field increased with depth. They concluded that soil could be a good repository for Sodium; however, the nutrient content of the resident saltbush population was affected. the background soil values for NPK were 21.58, 8.39 and 95.22 mg kg-1 at 0-20 cm soil depth while 19.54, 6.39 and 73.72 mg kg-1 at 20-40 cm soil depth, respectively. They concluded higher level of NPK in surface soils compared to sub-surface soils. Effluent irrigation affects pesticide sorption by altering: (i) the CEC of the soil; (ii) the soil solution pH which controls the dissociation and protonation of ionic pesticides as well as the number of available binding sites; (iii) the DOM content of the soil solution.

They concluded that there is considerable scope for scientists to help develop sustainable practices as well as more sound and rigorous effluent irrigation guidelines by improving the understanding of related soil processes. It was found increase in Electrical conductivity, Exchangeable cations, size and activity of soil microbial community. The irrigation with treated waste water improves soil labile Phosphorous content. Continuous application of effluent appears to deteriorate soil quality in some areas. They recommended the urgent management practices of waste water for irrigation purpose. Electrical Conductivity (salinity), organic carbon, soluble Ca and Mg, available soil phosphorus, potassium increased significantly and percentage of Total Neutralizing Value (T.N.V) of soil decreased. That physico chemical analysis of soil around industrial area and agricultural area shows variation. Plots irrigated with wastewater showed important structural damage in sub surface zone where soil pore network collapsed dramatically, resulting in a compact impermeable layer.

They concluded that a strong deterioration is observed in a very short time period due to such reasons. Poor water quality degrades soil quality, results in the accumulation of heavy metals and alteration of soil physical and chemical properties and influences the soil health to a great extent. They concluded the sewage effluents as a cheap potential alternative source of plant nutrients in agriculture. Finally sewage irrigation-it's never new, only recycle and reuse. The process of metal uptake, accumulation and distribution in plants is strongly influenced by the soil characteristics including pH, cation exchange capacity (CEC), organic matter, metals content, solubility sequences and plant species. The addition of organic and inorganic material would be expected to affect physicochemical changes on the soil, which would in turn impact upon soil microbial community structure, as has previously been shown following organic soil amendment.

The effluents of a leather industry complex markedly increased the concentrations of Cu, Fe, Mn, Zn, Al, Cr and Ni in sewage water, and irrigation with the contaminated sewage water increased the concentration of these elements in soils. Hence Mn, Cr and Ni are more likely to be the elements that may become health hazards for consumers of the crops grown in sewage-water irrigated fields. Under DFE irrigation there is a characteristic increase in EC, ESP, extractable soil Na, K and particularly P, and an increase in the size and activity of the soil microbial community. Despite these changes, the metabolic quotient and structure of the soil bacterial community remained unaffected suggesting that the existing soil microbial community is relatively resilient under dairy effluents irrigation. Treatment of industrial water at point source is the best remedy for consumers. Industrialists should take this responsibility.

Table: Analysis showing effects of effluent over soil characteristics at various places.

S. No.	Type of effluent used	Effect over soil	Reason
1	Urban effluent	Degradation	Accumulation of heavy metals like Cr, Pb, Cu, Ni, Zn and Cd.
2	Domestic sewage effluent	Fertilization	Improvement in organic matter, N, P, K and fertility of soil.
3	Dye, textile and wool industrial effluents	Degradation	Cd, Ni, Co accumulation.
4	Meatworks effluent	Fertilization	Decrease in bulk density and pH + increase in organic matter and nutrients.
5	Animal processing waste water	Degradation	Ca loss in drainage.

6	Sewage effluent	Degradation	High content of Zn, Cu, Fe, Ni, Pb but depletion in Mn.
7	Sewage effluent	Degradation	High amount of DTPA-Ni and heavy metals.
8	Secondary treated waste water	Degradation	Low salinity, high SAR, ESP values, low microbial content, accumulation of Cd, Ni and Cu.
9	Dairy factory effluents	Fertilization	Increase in Electrical conductivity, Exchangeable cations, size and activity of soil microbial community.
10	Phosphate fertilizer industry effluent	Fertilization	High amount of soil labile P content.
11	Paper mill wastewater	Fertilization	High amount alcohols, carbohydrates and inorganic compounds.
12	Olive mill waste water	Fertilization	High amount of N, P and K.
13	Municipal wastewater	Fertilization	Increase of EC, P, OM, TN, K, Na, Cl, Fe, Cd and Zn but decrease of soil pH.
14	Dye industrial effluent	Degradation	High amount of Chloride.
15	Winery wastewater	No significant effect	Low soil respiration, nitrogen cycling and microbial community structure.
16	Iron and steel industrial effluent	Degradation	High concentration of COD, BOD, Chloride content, Alkalanity and heavy metals like Ni, Cd, Pb, Fe and Mn.
17	Municipal wastewater	Fertilization	Increase in soil organic carbon and bulk density.
18	Municipal wastewater	Fertilization	Increase of Electrical Conductivity (salinity), organic carbon, soluble Ca and Mg, phosphorus and potassium.
19	Distillery irrigation	Fertilization	higher concentration of COD, BOD.
20	Industrial effluent	Fertilization	High organic C, Sulphate, Ca and Mg.
21	Vinasse effluent	Degradation	Nutrient leaching and soil dispersion, low ph.
22	Industrial effluent	Degradation	Collapsing of soil pore network.

References

- Harmful-effluents-industrial-discharge-pollution, harmful-effluents-industrial-discharge-pollution-2237: environmentalpollution.in, Retrieved 10 May, 2019
- Industrial-pollution: conservationinstitute.org, Retrieved 25 July, 2019
- Impact-of-industrial-wastes-on-aquatic-marine-and-fresh-water-life, waste-management- 42616: yourarti-clelibrary.com, Retrieved 13 August, 2019
- "Long-term impact of irrigation with sewage effluents on heavy metal content in soils, crops and groundwater—a case study", Agriculture, Ecosystems and Environment, 109, 310–322, 2005

Industrial Waste Management and Treatment

CHAPTER 4

The process of managing waste released by various industries is termed as industrial waste management. Its management and treatment methods include landfill, composting, aerobic and anaerobic treatment system, zero-sludge treatment, etc. This chapter closely examines these methods and related concepts of industrial waste management and treatment to provide an extensive understanding of the subject.

INDUSTRIAL WASTE MANAGEMENT

Industrial waste management is the set of strategies used to handle industrial waste removal and disposal in a timely but cost-effective manner that is also in compliance with regulations.

Importance of Industrial Waste Management

Good for the Environment

Poor industrial waste management can lead to those wastes getting into the local environment and then causing possibly irreparable damage. For example, chemical pollutants from manufacturing and similar industrial processes can enter the water and get into the bodies of everything living in that water. Certain chemicals can build up if ingested, meaning that higher and higher concentrations build up in the bodies of animals higher on the food chain, including potentially humans. Proper industrial waste management will help deter such outcomes while also ensuring that less hazardous industrial wastes are recycled, which is something that comes with its own benefits.

Protecting Human Health

Improper handling and disposal of industrial waste can contribute to increased incidents of serious medical conditions among the human residents of the local region. For example, if the run-off from a cattle ranch enters the local river, residents living down-river from that ranch can find themselves falling ill with diseases such as e-coli infections. Effective industrial waste management will not only do a better job of ensuring that such outcomes never happen, but also reducing the amount of industrial waste that is produced in the first place.

Legal Liability

Businesses have the obligation to effectively manage and dispose of their industrial waste. If they fail to do so, they will become liable for the damages caused through their negligence. Even in cases where the businesses win their cases, they will still lose out on legal fees and time spent defending themselves in the courts.

HAZARDOUS WASTE MANAGEMENT

Hazardous-waste management is the collection, treatment, and disposal of waste material that, when improperly handled, can cause substantial harm to human health and safety or to the environment. Hazardous wastes can take the form of solids, liquids, sludges, or contained gases, and they are generated primarily by chemical production, manufacturing, and other industrial activities. They may cause damage during inadequate storage, transportation, treatment, or disposal operations. Improper hazardous-waste storage or disposal frequently contaminates surface and groundwater supplies. People living in homes built near old and abandoned waste disposal sites may be in a particularly vulnerable position. In an effort to remedy existing problems and to prevent future harm from hazardous wastes, governments closely regulate the practice of hazardous-waste management.

Transport of Hazardous Waste

Hazardous waste generated at a particular site often requires transport to an approved treatment, storage, or disposal facility (TSDF). Because of potential threats to public safety and the environment, transport is given special attention by governmental agencies. In addition to the occasional accidental spill, hazardous waste has, in the past, been intentionally spilled or abandoned at random locations in a practice known as "midnight dumping." This practice has been greatly curtailed by the enactment of laws that require proper labeling, transport, and tracking of all hazardous wastes.

Transport Vehicles

Hazardous waste is generally transported by truck over public highways. Only a very small amount is transported by rail, and almost none is moved by air or inland waterway. Highway shipment is the most common because road vehicles can gain access to most industrial sites and approved TSDFs. Railroad trains require expensive siding facilities and are suitable only for very large waste shipments.

Hazardous wastes can be shipped in tank trucks made of steel or aluminum alloy, with capacities up to about 34,000 litres (9,000 gallons). They also can be containerized and shipped in 200-litre (55-gallon) drums. Specifications and standards for cargo tank trucks and shipping containers are included in governmental regulations.

Manifest System

In the United States a key feature of regulations pertaining to waste transport is the "cradle-to-grave" manifest system, which monitors the journey of hazardous waste from its point of origin to the point of final disposal. The manifest system helps to eliminate the problem of midnight dumping. It also provides a means for determining the type and quantity of hazardous waste being generated, as well as the recommended emergency procedures in case of an accidental spill. A manifest is a record-keeping document that must be prepared by the generator of the hazardous waste, such as a chemical manufacturer. The generator has primary responsibility for the ultimate disposal of the waste and must give the manifest, along with the waste itself, to a licensed waste transporter. A copy of the manifest must be delivered by the transporter to the recipient of the waste at an authorized TSDF. Each time the waste changes hands, a copy of the manifest must be signed. Copies of the manifest are kept by each party involved, and additional copies are sent to appropriate environmental agencies.

In the event of a leak or accidental spill of hazardous waste during its transport, the transporter must take immediate and appropriate actions, including notifying local authorities of the discharge. An area may have to be diked to contain the wastes, and efforts must be undertaken to remove the wastes and reduce environmental or public health hazards.

Treatment, Storage and Disposal

Several options are available for hazardous-waste management. The most desirable is to reduce the quantity of waste at its source or to recycle the materials for some other productive use. Nevertheless, while reduction and recycling are desirable options, they are not regarded as the final remedy to the problem of hazardous-waste disposal. There will always be a need for treatment and for storage or disposal of some amount of hazardous waste.

Treatment

Hazardous waste can be treated by chemical, thermal, biological, and physical methods. Chemical methods include ion exchange, precipitation, oxidation and reduction, and neutralization. Among thermal methods is high-temperature incineration, which not only can detoxify certain organic wastes but also can destroy them. Special types of thermal equipment are used for burning waste in either solid, liquid, or sludge form. These include the fluidized-bed incinerator, multiple-hearth furnace, rotary kiln, and liquid-injection incinerator. One problem posed by hazardous-waste incineration is the potential for air pollution.

Biological treatment of certain organic wastes, such as those from the petroleum industry, is also an option. One method used to treat hazardous waste biologically is called landfarming. In this technique the waste is carefully mixed with surface soil on a suitable tract of land. Microbes that can metabolize the waste may be added, along with nutrients. In some cases a genetically engineered species of bacteria is used. Food or forage crops are not grown on the same site. Microbes can also be used for stabilizing hazardous wastes on previously contaminated sites; in that case the process is called bioremediation.

The chemical, thermal, and biological treatment methods outlined above change the molecular form of the waste material. Physical treatment, on the other hand, concentrates, solidifies, or reduces the volume of the waste. Physical processes include evaporation, sedimentation, flotation, and filtration. Yet another process is solidification, which is achieved by encapsulating the waste in concrete, asphalt, or plastic. Encapsulation produces a solid mass of material that is resistant to leaching. Waste can also be mixed with lime, fly ash, and water to form a solid, cement-like product.

Surface Storage and Land Disposal

Hazardous wastes that are not destroyed by incineration or other chemical processes need to be disposed of properly. For most such wastes, land disposal is the ultimate destination, although it is not an attractive practice, because of the inherent environmental risks involved. Two basic methods of land disposal include landfilling and underground injection. Prior to land disposal, surface storage or containment systems are often employed as a temporary method.

Temporary on-site waste storage facilities include open waste piles and ponds or lagoons. New waste piles must be carefully constructed over an impervious base and must comply with regulatory requirements similar to those for landfills. The piles must be protected from wind dispersion or erosion. If leachate is generated, monitoring and control systems must be provided. Only non-containerized solid, non-flowing waste material can be stored in a new waste pile, and the material must be landfilled when the size of the pile becomes unmanageable.

A common type of temporary storage impoundment for hazardous liquid waste is an open pit or holding pond, called a lagoon. New lagoons must be lined with impervious clay soils and flexible membrane liners in order to protect groundwater. Leachate collection systems must be installed between the liners and groundwater monitoring wells are required. Except for some sedimentation, evaporation of volatile organics, and possibly some surface aeration, open lagoons provide no treatment of the waste. Accumulated sludge must be removed periodically and subjected to further handling as a hazardous waste.

Many older, unlined waste piles and lagoons are located above aquifers used for public water supply, thus posing significant risks to public health and environmental quality. A large number of these old sites have been identified and scheduled for cleanup, or remediation.

Secure Landfills

Landfilling of hazardous solid or containerized waste is regulated more stringently than landfilling of municipal solid waste. Hazardous wastes must be deposited in so-called secure landfills, which provide at least 3 metres (10 feet) of separation between the bottom of the landfill and the underlying bedrock or groundwater table. A secure hazardous-waste landfill must have two impermeable liners and leachate collection systems. The double leachate collection system consists of a network of perforated pipes placed above each liner. The upper system prevents the accumulation of leachate trapped in the fill, and the lower serves as a backup. Collected leachate is pumped to a treatment plant. In order to reduce the amount of leachate in the fill and minimize the potential for environmental damage, an impermeable cap or cover is placed over a finished landfill.

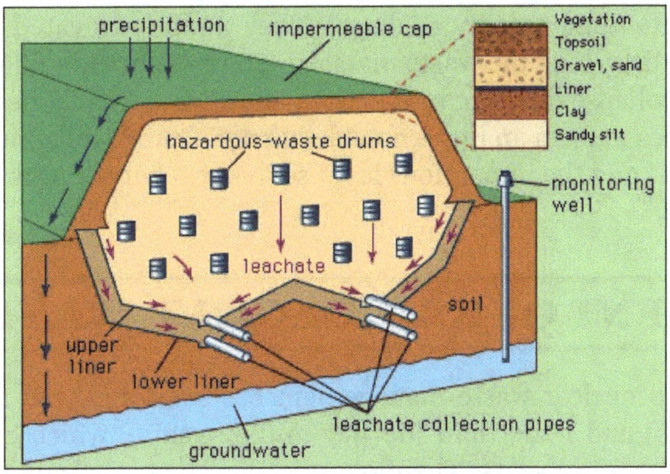

Schematic diagram of a secure hazardous-waste landfill with a double leachate collection system.

A groundwater monitoring system that includes a series of deep wells drilled in and around the site is also required. The wells allow a routine program of sampling and testing to detect any leaks or groundwater contamination. If a leak does occur, the wells can be pumped to intercept the polluted water and bring it to the surface for treatment.

One option for the disposal of liquid hazardous waste is deep-well injection, a procedure that involves pumping liquid waste through a steel casing into a porous layer of limestone or sandstone. High pressures are applied to force the liquid into the pores and fissures of the rock, where it is to be permanently stored. The injection zone must lie below a layer of impervious rock or clay, and it may extend more than 0.8 km (0.5 mile) below the surface. Deep-well injection is relatively inexpensive and requires little or no pretreatment of the waste, but it poses a danger of leaking hazardous waste and eventually polluting subsurface water supplies.

Remedial Action

Disposal of hazardous waste in unlined pits, ponds, or lagoons poses a threat to human health and environmental quality. Many such uncontrolled disposal sites were used in the past and have been abandoned. Depending on a determination of the level of risk, it may be necessary to remediate those sites. In some cases, the risk may require emergency action. In other instances, engineering studies may be required to assess the situation thoroughly before remedial action is undertaken.

One option for remediation is to completely remove all the waste material from the site and transport it to another location for treatment and proper disposal. This so-called off-site solution is usually the most expensive option. An alternative is on-site remediation, which reduces the production of leachate and lessens the chance of groundwater contamination. On-site remediation may include temporary removal of the hazardous waste, construction of a secure landfill on the same site, and proper replacement of the waste. It may also include treatment of any contaminated soil or groundwater. Treated soil may be replaced on-site and treated groundwater returned to the aquifer by deep-well injection.

A less costly alternative is full containment of the waste. This is done by placing an impermeable cover over the hazardous-waste site and by blocking the lateral flow of groundwater with subsurface cutoff walls. It is possible to use cutoff walls for this purpose when there is a natural layer of impervious soil or rock below the site. The walls are constructed around the perimeter of the site, deep enough to penetrate to the impervious layer. They can be excavated as trenches around the site without moving or disturbing the waste material. The trenches are filled with a bentonite clay slurry to prevent their collapse during construction, and they are backfilled with a mixture of soil and cement that solidifies to form an impermeable barrier. Cutoff walls thus serve as vertical barriers to the flow of water, and the impervious layer serves as a barrier at the bottom.

MANAGEMENT OF RADIOACTIVE WASTE

Of particular concern in nuclear waste management are two long-lived fission products, Tc-99 (half-life 220,000 years) and I-129 (half-life 15.7 million years), which dominate spent fuel radioactivity after a few thousand years. The most troublesome transuranic elements in spent fuel are Np-237 (half-life two million years) and Pu-239 (half-life 24,000 years). Nuclear waste requires sophisticated treatment and management to successfully isolate it from interacting with the biosphere. This usually necessitates treatment, followed by a long-term management strategy involving storage, disposal or transformation of the waste into a non-toxic form. Governments around the world are considering a range of waste management and disposal options, though there has been limited progress toward long-term waste management solutions.

In the second half of the 20th century, several methods of disposal of radioactive waste were investigated by nuclear nations, which are:

- "Long term above ground storage", not implemented.
- "Disposal in outer space" (for instance, inside the Sun), not implemented - as it would be currently too expensive.
- "Deep borehole disposal", not implemented.
- "Rock-melting", not implemented.
- "Disposal at subduction zones", not implemented.
- "Ocean disposal", used to be done by the USSR, the United Kingdom, Switzerland, the United States, Belgium, France, the Netherlands, Japan, Sweden, Russia, Germany, Italy and South Korea. This is no longer permitted by international agreements.
- "Sub seabed disposal", not implemented, not permitted by international agreements.
- "Disposal in ice sheets", rejected in Antarctic Treaty.
- "Direct injection", done by USSR and USA.
- Nuclear transmutation, using lasers to cause beta decay to convert the unstable atoms to those with shorter half-lives.

In the US, waste management policy completely broke down with the ending of work on the incomplete Yucca Mountain Repository. At present there are 70 nuclear power plant sites where spent fuel is stored. A Blue Ribbon Commission was appointed by President Obama to look into future options for this and future waste. A deep geological repository seems to be favored. 2018 Nobel Prize for Physics-winner Gérard Mourou, has proposed using Chirped pulse amplification to generate high-energy and low-duration laser pulses to transmute highly radioactive material (contained in a target) to significantly reduce its half-life, from thousands of years to only a few minutes.

Modern medium to high level transport container for nuclear waste.

Initial Treatment

Vitrification

The waste vitrification plant.

Long-term storage of radioactive waste requires the stabilization of the waste into a form which will neither react nor degrade for extended periods. It is theorized that one way to do this might be through vitrification. Currently at Sellafield the high-level waste (PUREX first cycle raffinate) is mixed with sugar and then calcined. Calcination involves passing the waste through a heated, rotating tube. The purposes of calcination are to evaporate the water from the waste, and de-nitrate the fission products to assist the stability of the glass produced.

The 'calcine' generated is fed continuously into an induction heated furnace with fragmented glass. The resulting glass is a new substance in which the waste products are bonded into the glass matrix when it solidifies. As a melt, this product is poured into stainless steel cylindrical containers ("cylinders") in a batch process. When cooled, the fluid solidifies ("vitrifies") into the glass. After being formed, the glass is highly resistant to water.

After filling a cylinder, a seal is welded onto the cylinder head. The cylinder is then washed. After being inspected for external contamination, the steel cylinder is stored, usually in an underground repository. In this form, the waste products are expected to be immobilized for thousands of years.

The glass inside a cylinder is usually a black glossy substance. All this work is done using hot cell systems. Sugar is added to control the ruthenium chemistry and to stop the formation of the volatile RuO_4 containing radioactive ruthenium isotopes. In the West, the glass is normally a borosilicate glass (similar to Pyrex), while in the former Soviet Union it is normal to use a phosphate glass. The amount of fission products in the glass must be limited because some (palladium, the other Pt group metals, and tellurium) tend to form metallic phases which separate from the glass. Bulk vitrification uses electrodes to melt soil and wastes, which are then buried underground. In Germany a vitrification plant is in use; this is treating the waste from a small demonstration reprocessing plant which has since been closed down.

Phosphate Ceramics

Vitrification is not the only way to stabilize the waste into a form that will not react or degrade for extended periods. Immobilization via direct incorporation into a phosphate based crystalline ceramic host is also used. The diverse chemistry of phosphate ceramics under various conditions demonstrate a versatile material that can withstand chemical, thermal and radioactive degradation over time. The properties of phosphates, particularly ceramic phosphates, of stability over a wide pH range, low porosity and minimization of secondary waste introduces possibilities for new waste immobilization techniques.

Ion Exchange

It is common for medium active wastes in the nuclear industry to be treated with ion exchange or other means to concentrate the radioactivity into a small volume. The much less radioactive bulk (after treatment) is often then discharged. For instance, it is possible to use a ferric hydroxide floc to remove radioactive metals from aqueous mixtures. After the radioisotopes are absorbed onto the ferric hydroxide, the resulting sludge can be placed in a metal drum before being mixed with cement to form a solid waste form. In order to get better long-term performance (mechanical stability) from such forms, they may be made from a mixture of fly ash, or blast furnace slag, and Portland cement, instead of normal concrete (made with Portland cement, gravel and sand).

Synroc

The Australian Synroc (synthetic rock) is a more sophisticated way to immobilize such waste, and this process may eventually come into commercial use for civil wastes (it is currently being developed for US military wastes). Synroc was invented by Prof Ted Ringwood (a geochemist) at the Australian National University. The Synroc contains pyrochlore and cryptomelane type minerals. The original form of Synroc (Synroc C) was designed for the liquid high level waste (PUREX raffinate) from a light water reactor. The main minerals in this Synroc are hollandite ($BaAl_2Ti_6O_{16}$), zirconolite ($CaZrTi_2O_7$) and perovskite ($CaTiO_3$). The zirconolite and perovskite are hosts for the actinides. The strontium and barium will be fixed in the perovskite. The caesium will be fixed in the hollandite.

Long Term Management

The time frame in question when dealing with radioactive waste ranges from 10,000 to 1,000,000 years, according to studies based on the effect of estimated radiation doses. Researchers suggest that forecasts of health detriment for such periods should be examined critically. Practical studies only consider up to 100 years as far as effective planning and cost evaluations are concerned. Long term behavior of radioactive wastes remains a subject for ongoing research projects in geoforecasting.

Above-ground Disposal

Dry cask storage typically involves taking waste from a spent fuel pool and sealing it (along with an inert gas) in a steel cylinder, which is placed in a concrete cylinder which acts as a radiation shield. It is a relatively inexpensive method which can be done at a central facility or adjacent to the source reactor. The waste can be easily retrieved for reprocessing.

Geologic Disposal

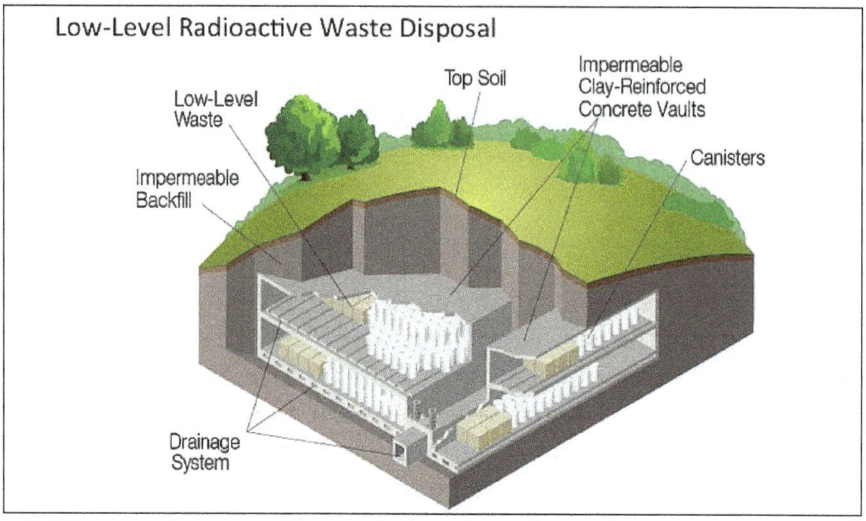

An underground low-level radioactive waste disposal site.

Radioactive materials at the Waste Isolation Pilot Plant leaked from a damaged storage drum due to the use of incorrect packing material. Analysis showed the lack of a "safety culture" at the plant since its successful operation for 15 years had bred complacency.

The process of selecting appropriate deep final repositories for high level waste and spent fuel is now under way in several countries with the first expected to be commissioned some time after 2010. The basic concept is to locate a large, stable geologic formation and use mining technology to excavate a tunnel, or large-bore tunnel boring machines (similar to those used to drill the Channel Tunnel from England to France) to drill a shaft 500 metres (1,600 ft) to 1,000 metres (3,300 ft) below the surface where rooms or vaults can be excavated for disposal of high-level radioactive waste. The goal is to permanently isolate nuclear waste from the human environment. Many people remain uncomfortable with the immediate stewardship cessation of this disposal system, suggesting perpetual management and monitoring would be more prudent.

Because some radioactive species have half-lives longer than one million years, even very low container leakage and radionuclide migration rates must be taken into account. Moreover, it may require more than one half-life until some nuclear materials lose enough radioactivity to cease being lethal to living things. A 1983 review of the Swedish radioactive waste disposal program by the National Academy of Sciences found that country's estimate of several hundred thousand years—perhaps up to one million years—being necessary for waste isolation "fully justified".

Ocean floor disposal of radioactive waste has been suggested by the finding that deep waters in the North Atlantic Ocean do not present an exchange with shallow waters for about 140 years based on oxygen content data recorded over a period of 25 years. They include burial beneath a stable abyssal plain, burial in a subduction zone that would slowly carry the waste downward into the Earth's mantle, and burial beneath a remote natural or human-made island. While these approaches all have merit and would facilitate an international solution to the problem of disposal of radioactive waste, they would require an amendment of the Law of the Sea.

Article 1 (Definitions) 7, of the 1996 Protocol to the Convention on the Prevention of Marine Pollution by Dumping of Wastes and Other Matter, (the London Dumping Convention) states:

> "Sea means all marine waters other than the internal waters of States, as well as the seabed and the subsoil thereof; it does not include sub-seabed repositories accessed only from land."

The proposed land-based subductive waste disposal method disposes of nuclear waste in a subduction zone accessed from land and therefore is not prohibited by international agreement. This method has been described as the most viable means of disposing of radioactive waste, and as the state-of-the-art as of 2001 in nuclear waste disposal technology. Another approach termed Remix and Return would blend high-level waste with uranium mine and mill tailings down to the level of the original radioactivity of the uranium ore, then replace it in inactive uranium mines. This approach has the merits of providing jobs for miners who would double as disposal staff, and of facilitating a cradle-to-grave cycle for radioactive materials, but would be inappropriate for spent reactor fuel in the absence of reprocessing, due to the presence of highly toxic radioactive elements such as plutonium within it.

Deep borehole disposal is the concept of disposing of high-level radioactive waste from nuclear reactors in extremely deep boreholes. Deep borehole disposal seeks to place the waste as much as 5 kilometres (3.1 mi) beneath the surface of the Earth and relies primarily on the immense natural geological barrier to confine the waste safely and permanently so that it should never pose a threat to the environment. The Earth's crust contains 120 trillion tons of thorium and 40 trillion tons of uranium (primarily at relatively trace concentrations of parts per million each adding up over the crust's $3 * 10^{19}$ ton mass), among other natural radioisotopes. Since the fraction of nuclides decaying per unit of time is inversely proportional to an isotope's half-life, the relative radioactivity of the lesser amount of human-produced radioisotopes (thousands of tons instead of trillions of tons) would diminish once the isotopes with far shorter half-lives than the bulk of natural radioisotopes decayed.

In January 2013, Cumbria county council rejected UK central government proposals to start work on an underground storage dump for nuclear waste near to the Lake District National Park. "For any host community, there will be a substantial community benefits package and worth hundreds of millions of pounds" said Ed Davey, Energy Secretary, but nonetheless, the local elected body voted 7–3 against research continuing, after hearing evidence from independent geologists that "the fractured strata of the county was impossible to entrust with such dangerous material and a hazard lasting millennia."

Transmutation

There have been proposals for reactors that consume nuclear waste and transmute it to other, less-harmful or shorter-lived, nuclear waste. In particular, the Integral Fast Reactor was a proposed nuclear reactor with a nuclear fuel cycle that produced no transuranic waste and in fact, could consume transuranic waste. It proceeded as far as large-scale tests, but was then canceled by the US Government. Another approach, considered safer but requiring more development, is to dedicate subcritical reactors to the transmutation of the left-over transuranic elements.

An isotope that is found in nuclear waste and that represents a concern in terms of proliferation is Pu-239. The large stock of plutonium is a result of its production inside uranium-fueled reactors and of the reprocessing of weapons-grade plutonium during the weapons program. An option for getting rid of this plutonium is to use it as a fuel in a traditional Light Water Reactor (LWR). Several fuel types with differing plutonium destruction efficiencies are under study.

Transmutation was banned in the US in April 1977 by President Carter due to the danger of plutonium proliferation, but President Reagan rescinded the ban in 1981. Due to the economic losses and risks, construction of reprocessing plants during this time did not resume. Due to high energy demand, work on the method has continued in the EU. This has resulted in a practical nuclear research reactor called Myrrha in which transmutation is possible. Additionally, a new research program called ACTINET has been started in the EU to make transmutation possible on a large, industrial scale. According to President Bush's Global Nuclear Energy Partnership (GNEP) of 2007, the US is now actively promoting research on transmutation technologies needed to markedly reduce the problem of nuclear waste treatment.

There have also been theoretical studies involving the use of fusion reactors as so called "actinide burners" where a fusion reactor plasma such as in a tokamak, could be "doped" with a small amount of the "minor" transuranic atoms which would be transmuted (meaning fissioned in the actinide case) to lighter elements upon their successive bombardment by the very high energy neutrons produced by the fusion of deuterium and tritium in the reactor. A study at MIT found that only 2 or 3 fusion reactors with parameters similar to that of the International Thermonuclear Experimental Reactor (ITER) could transmute the entire annual minor actinide production from all of the light water reactors presently operating in the United States fleet while simultaneously generating approximately 1 gigawatt of power from each reactor.

Re-use

Another option is to find applications for the isotopes in nuclear waste so as to re-use them. Already, caesium-137, strontium-90 and a few other isotopes are extracted for certain industrial applications such as food irradiation and radioisotope thermoelectric generators. While re-use does not eliminate the need to manage radioisotopes, it can reduce the quantity of waste produced.

The Nuclear Assisted Hydrocarbon Production Method, Canadian patent application 2,659,302, is a method for the temporary or permanent storage of nuclear waste materials comprising the placing of waste materials into one or more repositories or boreholes constructed into an unconventional oil formation. The thermal flux of the waste materials fracture the formation and alters the chemical and physical properties of hydrocarbon material within the subterranean formation to allow removal of the altered material. A mixture of hydrocarbons, hydrogen, and other formation fluids is produced from the formation. The radioactivity of high-level radioactive waste affords proliferation resistance to plutonium placed in the periphery of the repository or the deepest portion of a borehole.

Breeder reactors can run on U-238 and transuranic elements, which comprise the majority of spent fuel radioactivity in the 1,000–100,000-year time span.

Space Disposal

Space disposal is attractive because it removes nuclear waste from the planet. It has significant disadvantages, such as the potential for catastrophic failure of a launch vehicle, which could spread radioactive material into the atmosphere and around the world. A high number of launches would be required because no individual rocket would be able to carry very much of the material relative to the total amount that needs to be disposed of. This makes the proposal impractical economically and it increases the risk of at least one or more launch failures. To further complicate matters, international agreements on the regulation of such a program would need to be established. Costs and inadequate reliability of modern rocket launch systems for space disposal has been one of the motives for interest in non-rocket spacelaunch systems such as mass drivers, space elevators, and other proposals.

National Management Plans

Anti-nuclear protest near nuclear waste disposal centre.

Most countries are considerably ahead of the United States in developing plans for high-level radioactive waste disposal. Sweden and Finland are furthest along in committing to a particular disposal technology, while many others reprocess spent fuel or contract with France or Great Britain to do it, taking back the resulting plutonium and high-level waste. "An increasing backlog of plutonium from reprocessing is developing in many countries. It is doubtful that reprocessing makes economic sense in the present environment of cheap uranium".

In many European countries (e.g., Britain, Finland, the Netherlands, Sweden and Switzerland) the risk or dose limit for a member of the public exposed to radiation from a future high-level nuclear waste facility is considerably more stringent than that suggested by the International Commission on Radiation Protection or proposed in the United States. European limits are often more stringent than the standard suggested in 1990 by the International Commission on Radiation Protection by a factor of 20, and more stringent by a factor of ten than the standard proposed by the US Environmental Protection Agency (EPA) for Yucca Mountain nuclear waste repository for the first 10,000 years after closure.

The U.S. EPA's proposed standard for greater than 10,000 years is 250 times more permissive than the European limit. The U.S. EPA proposed a legal limit of a maximum of 3.5 millisieverts (350 millirem) each annually to local individuals after 10,000 years, which would be up to several percent of the exposure currently received by some populations in the highest natural background regions on Earth, though the U.S. DOE predicted that received dose would be much below that limit. Over a timeframe of thousands of years, after the most active short half-life radioisotopes decayed, burying U.S. nuclear waste would increase the radioactivity in the top 2000 feet of rock and soil in the United States (10 million km^2) by approximately 1 part in 10 million over the cumulative amount of natural radioisotopes in such a volume, but the vicinity of the site would have a far higher concentration of artificial radioisotopes underground than such an average.

Mongolia

After serious opposition had arisen about plans and negotiations between Mongolia with Japan and the United States of America to build nuclear-waste facilities in Mongolia, Mongolia stopped all negotiations in September 2011. These negotiations had started after U.S. Deputy Secretary of Energy Daniel Poneman visited Mongolia in September, 2010. Talks took place in Washington, D.C. between officials of Japan, the United States and Mongolia in February 2011. After this the United Arab Emirates (UAE), which wanted to buy nuclear fuel from Mongolia, joined in the negotiations. The talks were kept secret, and although news reported on them in May, Mongolia officially denied the existence of these negotiations. However, alarmed by this news, Mongolian citizens protested against the plans, and demanded the government withdraw the plans and disclose information. The Mongolian President Tsakhiagiin Elbegdorj issued a presidential order on September 13 banning all negotiations with foreign governments or international organizations on nuclear-waste storage plans in Mongolia. The Mongolian government has accused the newspaper of distributing false claims around the world. After the presidential order, the Mongolian president fired the individual who was supposedly involved in these conversations.

Illegal Dumping

Authorities in Italy are investigating a 'Ndrangheta mafia clan accused of trafficking and illegally dumping nuclear waste. According to a whistleblower, a manager of the Italy's state energy research agency Enea paid the clan to get rid of 600 drums of toxic and radioactive waste from Italy, Switzerland, France, Germany, and the US, with Somalia as the destination, where the waste was buried after buying off local politicians. Former employees of Enea are suspected of paying the criminals to take waste off their hands in the 1980s and 1990s. Shipments to Somalia continued into the 1990s, while the 'Ndrangheta clan also blew up shiploads of waste, including radioactive hospital waste, sending them to the sea bed off the Calabrian coast. According to the environmental group Legambiente, former members of the 'Ndrangheta have said that they were paid to sink ships with radioactive material for the last 20 years.

High-level Radioactive Waste Management

High-level radioactive waste management concerns how radioactive materials created during production of nuclear power and nuclear weapons are dealt with. Radioactive waste contains a mixture of short-lived and long-lived nuclides, as well as non-radioactive nuclides. There was reported some 47,000 tonnes of high-level nuclear waste stored in the USA in 2002.

The most troublesome transuranic elements in spent fuel are neptunium-237 (half-life two million years) and plutonium-239 (half-life 24,000 years). Consequently, high-level radioactive waste requires sophisticated treatment and management to successfully isolate it from the biosphere. This usually necessitates treatment, followed by a long-term management strategy involving permanent storage, disposal or transformation of the waste into a non-toxic form. Radioactive decay follows the half-life rule, which means that the rate of decay is inversely proportional to the duration of decay. In other words, the radiation from a long-lived isotope like iodine-129 will be much less intense than that of short-lived isotope like iodine-131.

Governments around the world are considering a range of waste management and disposal options, usually involving deep-geologic placement, although there has been limited progress toward implementing long-term waste management solutions. This is partly because the timeframes in question when dealing with radioactive waste range from 10,000 to millions of years, according to studies based on the effect of estimated radiation doses.

Thus, engineer and physicist Hannes Alfvén identified two fundamental prerequisites for effective management of high-level radioactive waste: (1) stable geological formations, and (2) stable human institutions over hundreds of thousands of years. As Alfvén suggests, no known human civilization has ever endured for so long, and no geologic formation of adequate size for a permanent radioactive waste repository has yet been discovered that has been stable for so long a period. Nevertheless, avoiding confronting the risks associated with managing radioactive wastes may create countervailing risks of greater magnitude. Radioactive waste management is an example of policy analysis that requires special attention to ethical concerns, examined in the light of uncertainty and futurity: consideration of 'the impacts of practices and technologies on future generations'.

There is a debate over what should constitute an acceptable scientific and engineering foundation for proceeding with radioactive waste disposal strategies. There are those who have argued, on the basis of complex geochemical simulation models, that relinquishing control over radioactive materials to geohydrologic processes at repository closure is an acceptable risk. They maintain that so-called "natural analogues" inhibit subterranean movement of radionuclides, making disposal of radioactive wastes in stable geologic formations unnecessary. However, existing models of these processes are empirically underdetermined: Due to the subterranean nature of such processes in solid geologic formations, the accuracy of computer simulation models has not been verified by empirical observation, certainly not over periods of time equivalent to the lethal half-lives of high-level radioactive waste. On the other hand, some insist deep geologic repositories in stable geologic formations are necessary. National management plans of various countries display a variety of approaches to resolving this debate.

Researchers suggest that forecasts of health detriment for such long periods should be examined critically. Practical studies only consider up to 100 years as far as effective planning and cost evaluations are concerned. Long term behaviour of radioactive wastes remains a subject for ongoing research. Management strategies and implementation plans of several representative national governments are described below.

Spent nuclear fuel stored underwater and uncapped.

Geologic Disposal

The International Panel on Fissile Materials has said:

> "It is widely accepted that spent nuclear fuel and high-level reprocessing and plutonium wastes require well-designed storage for periods ranging from tens of thousands to a million years, to minimize releases of the contained radioactivity into the environment. Safeguards are also required to ensure that neither plutonium nor highly enriched uranium is diverted to weapon use. There is general agreement that placing spent nuclear fuel in repositories hundreds of meters below the surface would be safer than indefinite storage of spent fuel on the surface".

The process of selecting appropriate permanent repositories for high level waste and spent fuel is now under way in several countries with the first expected to be commissioned some time after 2017. The basic concept is to locate a large, stable geologic formation and use mining technology to excavate a tunnel, or large-bore tunnel boring machines (similar to those used to drill the Channel Tunnel from England to France) to drill a shaft 500–1,000 meters below the surface where rooms or vaults can be excavated for disposal of high-level radioactive waste. The goal is to permanently isolate nuclear waste from the human environment. However, many people remain uncomfortable with the immediate stewardship cessation of this disposal system, suggesting perpetual management and monitoring would be more prudent.

Because some radioactive species have half-lives longer than one million years, even very low container leakage and radionuclide migration rates must be taken into account. Moreover, it may require more than one half-life until some nuclear materials lose enough radioactivity to no longer be lethal to living organisms. A 1983 review of the Swedish radioactive waste disposal program by the National Academy of Sciences found that country's estimate of several hundred thousand years—perhaps up to one million years—being necessary for waste isolation "fully justified."

The proposed land-based subductive waste disposal method would dispose of nuclear waste in a subduction zone accessed from land, and therefore is not prohibited by international agreement.

This method has been described as a viable means of disposing of radioactive waste, and as a state-of-the-art nuclear waste disposal technology.

In nature, sixteen repositories were discovered at the Oklo mine in Gabon where natural nuclear fission reactions took place 1.7 billion years ago. The fission products in these natural formations were found to have moved less than 10 ft (3 m) over this period, though the lack of movement may be due more to retention in the uraninite structure than to insolubility and sorption from moving ground water; uraninite crystals are better preserved here than those in spent fuel rods because of a less complete nuclear reaction, so that reaction products would be less accessible to groundwater attack.

Materials for Geological Disposal

In order to store the high level radioactive waste in long-term geological depositories, specific waste forms need to be used which will allow the radioactivity to decay away while the materials retain their integrity for thousands of years. The materials being used can be broken down into a few classes: glass waste forms, ceramic waste forms, and nanostructured materials.

The glass forms include borosilicate glasses and phosphate glasses. Borosilicate nuclear waste glasses are used on an industrial scale to immobilize high level radioactive waste in many countries which are producers of nuclear energy or have nuclear weaponry. The glass waste forms have the advantage of being able to accommodate a wide variety of waste-stream compositions, they are easy to scale up to industrial processing, and they are stable against thermal, radiative, and chemical perturbations. These glasses function by binding radioactive elements to nonradioactive glass-forming elements. Phosphate glasses while not being used industrially have much lower dissolution rates than borosilicate glasses, which make them a more favorable option. However, no single phosphate material has the ability to accommodate all of the radioactive products so phosphate storage requires more reprocessing to separate the waste into distinct fractions. Both glasses have to be processed at elevated temperatures making them unusable for some of the more volatile radiotoxic elements.

The ceramic waste forms offer higher waste loadings than the glass options because ceramics have crystalline structure. Also, mineral analogues of the ceramic waste forms provide evidence for long term durability. Due to this fact and the fact that they can be processed at lower temperatures, ceramics are often considered the next generation in high level radioactive waste forms. Ceramic waste forms offer great potential, but a lot of research remains to be done.

National Management Plans

Finland, the United States and Sweden are the most advanced in developing a deep repository for high-level radioactive waste disposal. Countries vary in their plans on disposing used fuel directly or after reprocessing, with France and Japan having an extensive commitment to reprocessing.

In many European countries (e.g., Britain, Finland, the Netherlands, Sweden and Switzerland) the risk or dose limit for a member of the public exposed to radiation from a future high-level nuclear waste facility is considerably more stringent than that suggested by the International Commission on Radiation Protection or proposed in the United States. European limits are often more stringent than the standard suggested in 1990 by the International Commission on Radiation Protection by a factor of 20, and more stringent by a factor of ten than the standard proposed by the U.S. Environmental Protection Agency (EPA) for Yucca Mountain nuclear waste repository for the first 10,000 years after closure. Moreover, the U.S. EPA's proposed standard for greater than 10,000 years is 250 times more permissive than the European limit.

The countries that have made the most progress towards a repository for high-level radioactive waste have typically started with public consultations and made voluntary siting a necessary condition. This consensus seeking approach is believed to have a greater chance of success than top-down modes of decision making, but the process is necessarily slow, and there is "inadequate experience around the world to know if it will succeed in all existing and aspiring nuclear nations".

Moreover, most communities do not want to host a nuclear waste repository as they are "concerned about their community becoming a de facto site for waste for thousands of years, the health and environmental consequences of an accident, and lower property values".

WASTE MANAGEMENT IN MINING INDUSTRY

The wastes being generated by the mining industry is mainly during the process of extraction, beneficiation and processing of minerals. Extraction which is the first phase that consists of initial removal of ore from the earthcrust which is done by the process of blasting which results in generation of large volume of waste such as soil, debris and other material which is useless for the industry and is just stored in big piles within the mine lease area, and sometimes, on public land. This is one of the way by which large amount of waste is generated by the process.

The bigger the scale of the mine, greater is the quantum of waste generated. Out of the two major types of mining methods (opencast and underground). The Opencast mining methods are therefore more pollution intensive as they generate 8 to 10 times more quantities of waste compared to the underground mines.

There is no estimation that how much waste is generated by the mining industry globally. But, everyone agrees that the quantity is so huge that is unimaginable. For example, the production of 1tonne of copper generates 110 tonnes of waste ore and 200 tones of overburden. Therefore it is a matter of concern that if the production of one metal is generating so much of waste, how much the entire industry would be generating.

Major Types of Waste Generated

Waste Rock

Mining operations generate two types of waste rock overburden and mine development rock.Overburden results from the development of surface mines, while mine development rock is a byproduct of mineral extraction in underground mines. The ratio of overburden excavated to the amount of mineral removed is called the overburden ratio or stripping ratio. For example a stripping ratio of 4:1 means that 4 tonnes of waste rock are removed to extract one tonne of ore.

As demand for coal increases to meet the country's energy requirement, the coal companies are digging deeper and deeper and even opting for lower grades of coal. The country is even planning for production from 300 m depths at stripping ratio of 1:15 for D and F grade quality of coal. If these mines were operational, it would mean that even if 1 million tonnes of coal were extracted, it would generate 15 million tonnes of waste material.

Tailings

This are an output of mineral beneficiation process. In this process the minerals which cannot be used for metal extraction directly has to be concentrated before it being used.

The entire process of concentration of the basic ore,is done by the grinding and milling process. Tailings are generally in the form of a slurry which contains certain hazardous contents such as arsenic, barite, calcite, cyanide, fluorite, mercury, pyrite and quartz.The slurry or the tailings is stored in a storage area called as the Tailings Dam or a Tailings Management Facility (TMF).

As this tailings are very hazardous both for the human beings and too for the environment so a proper disposal method of the tailings is to be done.

Tailings Disposal Method

Pond Storage

In this method the exhausted open pit mines are refilled with the tailings.

Dry Sacking

Here in this process the dewatering of tailings is done using vacuums and filters which save the water and reduces the impact on the environment.

Disposal into Underground Workings

In this process the disposal of the tailings is done in the exhausted underground mine which is truly a complex method.

Disposal into the Oceans

It is also known as STD (Submarine Tailings Disposal) or Deep Sea Tailings Disposal. Tailings can be conveyed using a pipeline then discharged so as to eventually descend into the depths. Practically, it is not an ideal method, as the close proximity to off-shelf depths is rare. STD is used, the depth of discharge is often what would be considered shallow, and extensive damage to the seafloor can result due to covering by the tailings product. It is also critical to control the density and temperature of the tailings product, to prevent it from travelling long distances or even floating to the surface.

Phytostabilisation

Phytostabilisation is a form of phytoremediation that uses plants for long-term stabilization and containment of tailings, by sequestering pollutants in soil near the roots. The plant's presence can reduce wind erosion, or the plant's roots can prevent water erosion, immobilize metals by adsorption or accumulation, and provide a zone around the roots where the metals can precipitate and stabilize. Pollutants become less bioavailable and livestock, wildlife, and human exposure is reduced. This approach can be especially useful in dry environments, which are subject to wind and water dispersion. Tailings are of great and growing concern in mining sector, specifically due to presence of heavy metals. The storage of tailings is commonly identified as the one of most important source of environmental impact for many mining operations. This is not surprising considering that the volume of tailings requiring storage can often exceed the in-situ total volume of the ore being mined and processed. In a single year, around 6.5 million tonnes of tailings is generated in the world.

Arsenic Contamination in Gold Mining

Elevated arsenic content is found in the surface and the ground water in the gold mining areas all round the world as the high content of arsenic in the surface as well as in the ground water is hazardous to the human beings and the several species who are the habitat of that particular mining area. So an extensive hydrological area is compulsory to be done.

Measure to be taken regarding arsenic contamination:

- More concentration should be given to the potential impacts of spent ore and waste rock disposal on the surface and ground water.
- Many natural soils and man-made clay liners have the capacity to minimize the content of arsenic in the mine water.

Overburden Management

As a large amount of overburden is being generated by the mining industry which is a great threat to the environment, therefore a perfect well management of this overburden should be done. The best use for overburden waste is to backfill the excavated land.

Some of the most common method of stabilization is by plantation. Overburden dumps are generally reclaimed by tree species as plantation improves the moisture contents, bulk density, pH and overall nutrient contents of soils. Tree species like Dalbergia sissoo, Eucalyptus, Cassia seamea, Acaccia mangium and Peltaphorum are ideal for bioreclamation of overburden dumps. some other preventive measure to be taken at the initial stage are.

- Excavation from a new pit should begin only after an existing pit has been exhausted.
- Till a pit is exhausted, the overburden should be properly compacted and stacked in. Specified locations in low-lying non-mineralized zones within the lease area.
- The height and slope of the overburden dumps should be maintained to prevent accidents.
- Drainage should be considered to handle heavy rainfall.
- Sedimentation.

PACKAGING WASTE MANAGEMENT

Packaging Design Requirements

Minimise the volume and weight of packaging you use. The amount of packaging must be the minimum required to ensure the safety and hygiene of the packed product and to be acceptable to the consumer.

Make sure packaging has a minimal environmental impact after it is disposed of.

You must ensure packaging does not contain high levels of noxious or hazardous substances. Make sure the amount of heavy metals (cadmium, mercury, lead and hexavalent chromium, or any combination of these) does not exceed 100 parts per million by weight. These limits apply to packaging plus any packaging components, calculated as a whole unit. For example, components such as lids do not have to meet these limits independently. There are a few exemptions from the limits.

Make sure your packaging can be reused or recovered through recycling, energy recovery or composting.

Packaging Design for Reuse and Recovery

Packaging you intend to be recovered by recycling must be manufactured so that a percentage (by weight) of the material can be recycled. The percentage varies according to the type of material and the current British standards.

Packaging you intend to be recovered by energy recovery must be processed to allow for the maximum amount of energy to be recovered. This means that if you burn any packaging waste, it must produce more energy than is used by the incineration process.

Packaging should contain at least 50 per cent of combustible organic materials by weight. Combustible organic materials include paper, wood, cardboard and other organic fibres.

Packaging intended for recovery by composting must be biodegradable so it doesn't slow down the composting process. To be biodegradable the packaging must decompose into carbon dioxide, biomass and water.

Reusable packaging must be capable of being used several times. Once at the end of its useful life, it must also meet the requirements for recycling, energy recovery or composting.

The best way to demonstrate you comply with the packaging design requirements is to follow the European Standards on designing packaging.

Keep Records of Packaging

Keep evidence that your packaging complies with the Packaging (Essential Requirements) Regulations for at least four years from the date the packaging was first placed on the market. You need to be able to supply these details within 28 days, if requested.

How you hold this information is up to you. You may need to keep a variety of document types to ensure you have evidence for all of the requirements.

You could keep records of:

- Existing technical documents on the design and use of your packaging.
- Existing quality assurance documents.
- Environmental management systems and auditing documents.
- Documents relating to other packaging standards.

Reprocessing and Exporting Packaging Waste

If you recover, recycle or export packaging waste, you can apply for accreditation. This will allow you to issue electronic packaging waste recovery notes (ePRNs) and electronic packaging waste export recovery notes (ePERNs) to obligated businesses and compliance schemes.

ePRNs and ePERNs can be issued for each tonne of packaging waste reprocessed or exported for reprocessing. Businesses and compliance schemes obtain ePRNs or ePERNs to show that they have met their packaging recovery and recycling obligation. As an accredited reprocessor or exporter you can charge for this service.

Even if the packaging regulations do not apply to you, it is a good idea to reduce, reuse and recycle packaging. This can save your business money and reduce your waste disposal costs.

Buying Packaging and Packaged Products

Create guidelines for packaging for your staff to follow when buying goods for your business.

Buy goods in bulk to reduce packaging. For example:

- Buy paint in one 5-litre drum, instead of buying five individual 1-litre drums.
- Buy bigger jars of coffee, instead of buying smaller jars frequently.

You could also:

- Remove disposable plastic cups from drinks machines, and ask staff to use their own mugs instead.
- Use bowls of sugar, jugs of milk and reusable or biodegradable cutlery, when organising events, meetings or conferences, instead of many small, individually packaged products.
- Buy recycled office stationery that uses recyclable or biodegradable packaging.

Selling Packaging

If you sell packaging, tell your customers about the return, recovery and collection facilities available to them.

Use recyclable or biodegradable materials to package your products.

Reusing Packaging

You could:

- Introduce returnable and reusable packaging in co-operation with your supply chain and customers.
- Shred old paper to reuse as packaging filling.
- Keep old bubble wrap or polystyrene to reuse as packaging.
- Reuse packaging such as wooden pallets, cardboard boxes or jiffy bags.

Disposing of Packaging

Reduce, reuse and recycle your packaging waste. Separate different types of packaging for recycling - this should reduce your recycling costs.

Cover any waste packaging stored in skips to prevent pollution and to minimise wind-blown dust and litter.

If packaging is intended for reuse, make sure you have documented the systems for supply, return, reuse and refill.

If there is a risk that the packaging could have been contaminated during its use, you must ensure it is kept dry and that any run-off doesn't pollute water or land. Ensure your site drainage system is connected to the sewer system.

Buy a compactor or baler, which crushes packaging into blocks. This will allow you to transport larger volumes of packaging, which will reduce your recycling or disposal costs.

WASTE MANAGEMENT IN FOOD INDUSTRY

Food processing industry around the world is making serious efforts to minimize by-products, compost organic waste, recycle processing and packaging materials, and save energy and water. The three R's of waste management – Reduce, Reuse and Recycle – can help food manufacturers in reducing the amount of waste sent to landfill and reusing waste.

Food Recovery Hierarchy

EPA's Food Recovery Hierarchy is an excellent resource to follow for food processors and beverage producers as it provides the guidance to start a program that will provide the most benefits for the environment, society and the food manufacturer.

Notably, landfill is the least favored disposal option for waste generated in food and beverage producers worldwide. There are sustainable, effective and profitable waste management options including:

- Making animal feed.
- Composting to create nutrient-rich fertilizer.
- Anaerobic digestion to produce energy-rich biogas.
- Recycling/reusing waste for utilization by other industries.
- Feeding surplus food to needy people.

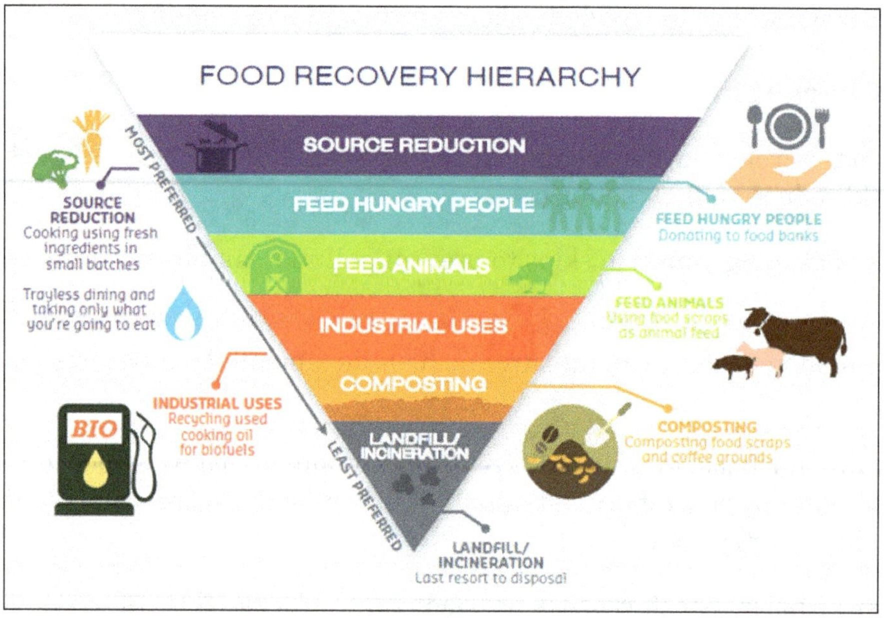

Waste Management Options

Food manufacturers has a unique problem – excess product usually has a relatively short shelf life while most of the waste is organic in nature. Food waste created during the production process can be turned into animal feed and sold to goat farms, chicken farms etc. As far as WWTP sludge is concerned, top food manufacturers are recycling/reusing it through land application, anaerobic digestion and composting alternatives.

Organic waste at any food processing plant can be composted in a modern in-vessel composting and the resultant fertilizer can be used for in-house landscaping or sold as organic fertilizer as attractive prices.

Another plausible way of managing organic waste at the food manufacturing plant is to biologically degrade it in an anaerobic digester leading to the formation of energy-rich biogas and digestate. Biogas can be used as a heating fuel in the plant itself or converted into electricity by using a CHP unit while digestate can be used as a soil conditioner. Biogas can also be converted into biomethane or bio-CNG for its use as vehicle fuel.

Items such as cardboard, clean plastic, metal and paper are all commodities that can be sold to recyclers. Lots of cardboard boxes are used by food manufacturers for supplies which can be broken down into flat pieces and sold to recyclers.

Cardboard boxes can also be reused to temporarily store chip packages before putting them into retail distribution boxes. Packaging can be separated in-house and recovered using "jet shredder" waste technologies which separate film, carton and foodstuffs, all of which can then be recycled separately.

Organizing a Zero-landfill Program

How do you develop a plan to create a zero-landfill or zero waste program in food and beverage producing company? The best way to begin is to start at a small-level and doing what you can. Perfect those programs and set goals each year to improve. Creation of a core team is an essential step in order to explore different ways to reduce waste, energy and utilities.

Measuring different waste streams and setting a benchmark is the initial step in the zero-landfill program. Once the data has been collected, we should break these numbers down into categories, according to the EPA's Food Recovery Challenge and identify the potential opportunities.

For example, inorganic materials can be categorized based on their end lives (reuse, recycle or landfill). The food and beverage industry should perform a waste sort exercise (or dumpster dive) to identify its key streams.

INDUSTRIAL WASTE TREATMENT

Many different types of solid, liquid, and gaseous wastes are discharged by industries. Most industrial waste is recycled, treated and discharged, or placed in a landfill. There is no one means of managing industrial wastes because the nature of the wastes varies widely from one industry to another. One company might generate a waste that can be treated readily and discharged to the environment (direct discharge) ortoa sewer in which case final treatment might be accomplished at a publicly owned treatment works (POTW). Treatment at the company before discharge to a sewer is referred to as pretreatment. Another company might generate a waste which is regarded as hazardous and therefore requires special management procedures related to storage, transportation and final disposal.

The pertinent legislation governing to what extent wastewaters need to be treated before discharge is the 1972 Clean Water Act (CWA). Major amendments to the CWA were passed in 1977 and 1987. The Environmental Protection Agency (EPA) was also charged with the responsibility of regulating the priority pollutants under the CWA. The CWA specifies that toxic and nonconventional pollutants are to be treated with the Best Available Technology (BAT). Gaseous pollutants are regulated under the Clean Air Act (CAA), promulgated in 1970 and amended in 1977 and 1990. An important part of the CAA consists of measures to attain and maintain National Ambient Air Quality Standards (NAAQS). Hazardous air pollutant (HAP) emissions are to be controlled through Maximum Achievable Control Technology (MACT) which can include process changes, material substitutions and air pollution control equipment. The "cradle to grave" management of hazardous wastes is to be performed in accordance with the Resource Conservation and Recovery Act (RCRA) of 1976 and the Hazardous and Solid Waste Amendments (HSWA) of 1984.

In 1990, the United States, through the Pollution Prevention Act, adopted a program designed to reduce the volume and toxicity of waste discharges. Pollution prevention (P2) strategies might involve changing process equipment or chemistry, developing new processes, eliminating products, minimizing wastes, recycling water or chemicals, trading wastes with another company, etc. In 1991, the EPA instituted the 33/50 program which was to result in an overall 33% reduction of 17 high priority pollutants by 1992 and a 50% reduction of the pollutants by 1995. Both goals were surpassed. Not only has this program been successful, but it sets an important precedence because the participating companies volunteered. Additionally, P2 efforts have led industries to rigorously think through product life cycles. A Life Cycle Analysis (LCA) starts with consideration for acquiring raw materials, moves through the stages related to processing, assembly, service and reuse, and ends with retirement/disposal. The LCA therefore reveals to industry the costs and problems versus the benefits for every stage in the life of a product.

In designing a waste management program for an industry, one must think first in terms of P2 opportunities, identify and characterize the various solid, liquid and gaseous waste streams, consider relevant legislation, and then design an appropriate waste management system. Treatment systems that rely on physical (e.g., settling, floatation, screening, sorption, membrane technologies, air stripping) and chemical (e.g., coagulation, precipitation, chemical oxidation and reduction, pH adjustment) operations are referred to as physicochemical, whereas systems in which microbes are cultured to metabolize waste constituents are known as biological processes (e.g., activated sludge, trickling filters, biotowers, aerated lagoons, anaerobic digestion, aerobic digestion, composting). Oftentimes, both physicochemical and biological systems are used to treat solid and liquid waste streams. Biological systems might be used to treat certain gas streams, but most waste gas streams are treated physicochemically (e.g., cyclones, electrostatic precipitators, scrubbers, bag filters, thermal methods). Solids and the sludges or residuals that result from treating the liquid and gaseous waste streams are also treated by means of physical, chemical, and biological methods.

In many cases, the systems used to treat wastes from domestic sources are also used to treat industrial wastes. For example, municipal wastewaters often consist of both domestic and industrial waste. The local POTW therefore may be treating both types of wastes. To avoid potential problems caused by the input of industrial wastes, municipalities commonly have pretreatment programs which require that industrial wastes discharged to the sewer meet certain standards. The standards generally include limits for various toxic agents such as metals, organic matter measured in terms of biochemical oxygen demand (bod) or chemical oxygen demand, nutrients such as nitrogen and phosphorus, pH and other contaminants that are recognized as having the potential to impact on the performance of the POTW. At the other end of the spectrum, there are wastes that need to be segregated and managed separately in special systems. For example, an industry might generate a hazardous waste that needs to be placed in barrels and transported to an EPA approved treatment, storage or disposal facility (TSDF).

Thus, it is not possible to simply use one train of treatment operations for all industrial waste streams, but an effective, generic strategy has been developed in recent years for considering the waste management options available to an industry. The basis for the strategy is to look for P2 opportunities and to consider the life cycle of a product. An awareness of waste stream characteristics and the potential benefits of stream segregation is then melded with the knowledge of regulatory compliance issues and treatment system capabilities/performance to minimize environmental risks and costs.

METHODS OF INDUSTRIAL WASTE TREATMENT

Waste Sorting

Waste sorting is the process by which waste is separated into different elements. Waste sorting can occur manually at the household and collected through curbside collection schemes, or automatically separated in materials recovery facilities or mechanical biological treatment systems. Hand sorting was the first method used in the history of waste sorting.

Waste can also be sorted in a civic amenity site.

"Waste segregation" means dividing waste into dry and wet. Dry waste includes wood and related products, metals and glass. Wet waste typically refers to organic waste usually generated by eating establishments and are heavy in weight due to dampness. Waste can also be segregeconomic concern.

Recycling bins.

Manual waste sorting for recycling.

Methods

Waste is collected at its source in each area and separated. The way that waste is sorted must re-flect local disposal systems. The following categories are common:

- Paper.
- Cardboard (including packaging for return to suppliers).
- Glass (clear, tinted – no light bulbs or window panes, which belong with residual waste).
- Plastics.
- Textiles.
- Wood, leather, rubber.
- Scrap metal.
- Compost.
- Special/hazardous waste.
- Residual waste.

Organic waste can also be segregated for disposal:

- Leftover food which has had any contact with meat can be collected separately to prevent the spread of bacteria:
    - Meat and bone can be retrieved by bodies responsible for animal waste.
    - If other leftovers are sent, for example, to local farmers, they can be sterilised before being fed to the animals.
- Peels and scrapings from fruit and vegetables can be composted along with other degrad-able matter. Other waste can be included for composting, such as cut flowers, corks, coffee grounds, rotting fruit, tea bags, eggshells and nutshells, and paper towels.

Chip pan oil (fryer oil), used fats, vegetable oil and the content of fat filters can be collected by companies able to re-use them. Local authority waste departments can provide relevant addresses.

This can be achieved by providing recycling bins.

Characteristic containers for recycling.

Garbage containers.

Landfill

A landfill site (also known as a tip, dump, rubbish dump, garbage dump or dumping ground and historically as a midden) is a site for the disposal of waste materials by burial. Landfill is the oldest form of waste treatment, although the burial of the waste is modern; historically, refuse was simply left in piles or thrown into pits. Historically, landfills have been the most common method of organized waste disposal and remain so in many places around the world.

A landfill.

Some landfills are also used for waste management purposes, such as the temporary storage, consolidation and transfer, or processing of waste material (sorting, treatment, or recycling). Unless they are stabilized, these areas may experience severe shaking or soil liquefaction of the ground during a large earthquake.

One of several landfills.

Operations

Typically, operators of well-run landfills for non-hazardous waste meet predefined specifications by applying techniques to:

- Confine waste to as small an area as possible.
- Compact waste to reduce volume.

They can also cover waste (usually daily) with layers of soil or other types of material such as woodchips and fine particles.

During landfill operations, a scale or weighbridge may weigh waste collection vehicles on arrival and personnel may inspect loads for wastes that do not accord with the landfill's waste-acceptance criteria. Afterward, the waste collection vehicles use the existing road network on their way to the tipping face or working front, where they unload their contents. After loads are deposited, compactors or bulldozers can spread and compact the waste on the working face. Before leaving the landfill boundaries, the waste collection vehicles may pass through a wheel-cleaning facility. If necessary, they return to the weighbridge for re-weighing without their load. The weighing process can assemble statistics on the daily incoming waste tonnage, which databases can retain for record keeping. In addition to trucks, some landfills may have equipment to handle railroad containers. The use of "rail-haul" permits landfills to be located at more remote sites, without the problems associated with many truck trips.

Typically, in the working face, the compacted waste is covered with soil or alternative materials daily. Alternative waste-cover materials include chipped wood or other "green waste", several sprayed-on foam products, chemically "fixed" bio-solids, and temporary blankets. Blankets can be lifted into place at night and then removed the following day prior to waste placement. The space that is occupied daily by the compacted waste and the cover material is called a daily cell. Waste compaction is critical to extending the life of the landfill. Factors such as waste compressibility, waste-layer thickness and the number of passes of the compactor over the waste affect the waste densities.

Sanitary Landfill Life Cycle

The term landfill is usually shorthand for a municipal landfill or sanitary landfill. These facilities were first introduced early in the 20th century, but gained wide use in the 1960s and 1970s, in an effort to eliminate open dumps and other "unsanitary" waste disposal practices. The sanitary landfill is an engineered facility that separates and confines waste. Sanitary landfills are intended as biological reactors (bioreactors) in which microbes will break down complex organic waste into simpler, less toxic compounds over time. These reactors must be designed and operated according to regulatory standards and guidelines.

Usually, aerobic decomposition is the first stage by which wastes are broken down in a landfill. These are followed by four stages of anaerobic degradation. Usually, solid organic material in solid phase decays rapidly as larger organic molecules degrade into smaller molecules. These smaller organic molecules begin to dissolve and move to the liquid phase, followed by hydrolysis of these organic molecules, and the hydrolyzed compounds then undergo transformation and volatilization as carbon dioxide (CO_2) and methane (CH_4), with rest of the waste remaining in solid and liquid phases.

During the early phases, little material volume reaches the leachate, as the biodegradable organic matter of the waste undergoes a rapid decrease in volume. Meanwhile, the leachate's chemical oxygen demand increases with increasing concentrations of the more recalcitrant compounds compared to the more reactive compounds in the leachate. Successful conversion and stabilization of the waste depends on how well microbial populations function in syntrophy, i.e. an interaction of different populations to provide each other's nutritional needs.

The life cycle of a municipal landfill undergoes five distinct phases:

Phase I: Initial Adjustment

As the waste is placed in the landfill, the void spaces contain high volumes of molecular oxygen (O_2). With added and compacted wastes, the O_2 content of the landfill bioreactor strata gradually decreases. Microbial populations grow, density increases. Aerobic biodegradation dominates, i.e. the primary electron acceptor is O_2.

Phase II: Transition

The O_2 is rapidly degraded by the existing microbial populations. The decreasing O_2 leads to less aerobic and more anaerobic conditions in the layers. The primary electron acceptors during transition are nitrates and sulphates, since O_2 is rapidly displaced by CO_2 in the effluent gas.

Phase III: Acid Formation

Hydrolysis of the biodegradable fraction of the solid waste begins in the acid formation phase, which leads to rapid accumulation of volatile fatty acids (VFAs) in the leachate. The increased organic acid content decreases the leachate pH from approximately 7.5 to 5.6. During this phase, the decomposition intermediate compounds like the VFAs contribute much chemical oxygen demand (COD). Long-chain volatile organic acids (VOAs) are converted to acetic acid ($C_2H_4O_2$), CO_2, and hydrogen gas (H_2). High concentrations of VFAs increase both the biochemical oxygen demand (BOD) and VOA concentrations, which initiates H_2 production by fermentative bacteria, which stimulates the growth of H_2-oxidizing bacteria. The H_2 generation phase is relatively short because it is complete by the end of the acid formation phase. The increase in the biomass of acidogenic bacteria increases the amount of degradation of the waste material and consuming nutrients. Metals, which are generally more water-soluble at lower pH, may become more mobile during this phase, leading to increasing metal concentrations in the leachate.

Phase IV: Methane Fermentation

The acid formation phase intermediary products (e.g. acetic, propionic, and butyric acids) are converted to CH_4 and CO_2 by methanogenic microorganisms. As VFAs are metabolized by the methanogens, the landfill water pH returns to neutrality. The leachate's organic strength, experessed as oxygen demand, decreases at a rapid rate with increases in CH_4 and CO_2 gas production. This is the longest decomposition phase.

Phase V: Final Maturation and Stabilization

The rate of microbiological activity slows during the last phase of waste decomposition as the supply of nutrients limits the chemical reactions, e.g. as bioavailable phosphorus becomes increasingly scarce. CH_4 production almost completely disappears, with O_2 and oxidized species gradually reappearing in the gas wells as O_2 permeates downwardly from the troposphere. This transforms the oxidation–reduction potential (ORP) in the leachate toward oxidative processes. The residual organic materials may incrementally be converted to the gas phase, and as organic matter is composted; i.e. the organic matter is converted to humic-like compounds.

Social and Environmental Impact

Landfill operation. Note that the area being filled is a single, well-defined "cell" and that a protective landfill liner is in place (exposed on the left) to prevent contamination by leachates migrating downward through the underlying geological formation.

Landfills have the potential to cause a number of issues. Infrastructure disruption, such as damage to access roads by heavy vehicles, may occur. Pollution of local roads and water courses from wheels on vehicles when they leave the landfill can be significant and can be mitigated by wheel washing systems. Pollution of the local environment, such as contamination of groundwater or aquifers or soil contamination may occur, as well.

Leachate

When precipitation falls on open landfills, water percolates through the garbage and becomes contaminated with suspended and dissolved material, forming leachate. If this is not contained it can contaminate groundwater. All modern landfill sites use a combination of impermeable liners several metres thick, geologically stable sites and collection systems to contain and capture this leachate. It can then be treated and evaporated. Once a landfill site is full, it is sealed off to prevent precipitation ingress and new leachate formation. However, liners must have a lifespan, be it several hundred years or more. Eventually, any landfill liner could leak, so the ground around landfills must be tested for leachate to prevent pollutants from contaminating groundwater.

Decomposition Gases

Rotting food and other decaying organic waste creates decomposition gases, especially CO_2 and CH_4 from aerobic and anaerobic decomposition, respectively. Both processes occur simultaneously in different parts of a landfill. In addition to available O_2, the fraction of gas constituents will vary, depending on the age of landfill, type of waste, moisture content and other factors. For example, the maximum amount of landfill gas produced can be illustrated a simplified net reaction of diethyl oxalate that accounts for these simultaneous reactions:

$$C_6H_{10}O_4 + 1{:}5H_2O \rightarrow 3{:}25CH_4 + 2{:}75CO_2$$

On average, about half of the volumetric concentration of landfill gas is CH_4 and slightly less than half is CO_2. The gas also contains about 5% molecular nitrogen (N_2), less than 1% hydrogen sulfide (H_2S), and a low concentration of non-methane organic compounds (NMOC), about 2700 ppmv.

Landfill gases can seep out of the landfill and into the surrounding air and soil. Methane is a greenhouse gas, and is flammable and potentially explosive at certain concentrations, which makes it perfect for burning to generate electricity cleanly. Since decomposing plant matter and food waste only release carbon that has been captured from the atmosphere through photosynthesis, no new carbon enters the carbon cycle and the atmospheric concentration of CO_2 is not affected. Carbon dioxide traps heat in the atmosphere, contributing to climate change. In properly managed landfills, gas is collected and flared or recovered for landfill gas utilization.

Vectors

Poorly run landfills may become nuisances because of vectors such as rats and flies which can spread infectious diseases. The occurrence of such vectors can be mitigated through the use of daily cover.

Other Nuisances

Other potential issues include wildlife disruption, dust, odor, noise pollution, and reduced local property values.

Landfill Gas

Gases are produced in landfills due to the anaerobic digestion by microbes. In a properly managed landfill this gas is collected and used. Its uses range from simple flaring to the landfill gas utilization and generation of electricity. Landfill gas monitoring alerts workers to the presence of a build-up of gases to a harmful level. In some countries, landfill gas recovery is extensive; in the United States, for example, more than 850 landfills have active landfill gas recovery systems.

A gas flare produced by a landfill.

Types of Landfill Gas

- Municipal solid waste: Takes in household waste and nonhazardous material. Included in this type of landfill is a Bioreactor Landfill that specifically degrades organic material.
- Industrial waste: For commercial and industrial waste. Other related landfills include Construction and Demolition Debris Landfills and Coal Combustion Residual Landfills.
- Hazardous waste or PCB waste: Polychlorinated Biphenyl (PCB) landfills that are monitored in the United States by the Toxic Substances Control Act of 1976 (TSCA).

Microbial Topics

The status of a landfill's microbial community may determine its digestive efficiency.Bacteria that digest plastic have been found in landfills.

Reclaiming Materials

One can treat landfills as a viable and abundant source of materials and energy. In the developing world, waste pickers often scavenge for still-usable materials. In commercial contexts, companies have also discovered landfill sites, and many have begun harvesting materials and energy. Well-known examples include gas-recovery facilities. Other commercial facilities include waste incinerators which have built-in material recovery. This material recovery is possible through the use of filters (electro filter, active-carbon and potassium filter, quench, HCl-washer, SO_2-washer, bottom ash-grating, etc.).

Alternatives

In addition to waste reduction and recycling strategies, there are various alternatives to landfills, including waste-to-energy incineration, anaerobic digestion, composting, mechanical biological treatment, pyrolysis and plasma arc gasification. Depending on local economics and incentives, these can be made more financially attractive than landfills.

Composting

Composting is the biological decomposition of organic waste such as food or plant material by bacteria, fungi, worms and other organisms under controlled aerobic (occurring in the presence of oxygen) conditions. The end result of composting is an accumulation of partially decayed organic matter called humus. Composting with worms, also known as vermiculture, results in nutrient- loaded worm castings.

- It's easy.
- It creates a useful soil enricher.
- It is an environmentally sound way of reducing yard waste.

Yard waste is such materials as leaves, grass clippings, brush, and prunings. Some states, including Florida, have banned yard waste from landfills. Leaves and grass clippings can be used as mulch in your garden or landscape. Yard waste that will be picked up should be bundled or bagged.

Step 1: Choose the right composting method for you.

There are two kinds of composting-bin/pile composting and worm composting (vermiculture). The type of composter used should be the one that best suits your needs and capabilities.

Bin/Pile Composting	Worm Composting
Simply throw in organic materials as they become available around your home and yard.	Popular way to compost small amounts of food and paper wastes.
Learn about the right bin for you.	Read detailed instructions on how to build a worm bin.

Step 2: Choose the right location for your compost bin.

Consider how you will get the raw materials to the pile and how the finished compost will be moved to the area it will be used.

Step 3: Decide what to compost.

To build a compost pile, simply alternate layers of browns and greens.

Greens:

- Vegetable and food scraps.
- Fresh grass clippings and yard waste.
- Coffee gounds.

- Tea bags.
- Egg shells.

Browns:

- Dried leaves, grass, mulch or hay.
- Cardboard rolls.
- Sawdust.
- Lint.
- Newspaper (shredded).

Step 4: The composting process.

The compost pile should be periodically mixed to incorporate oxygen. Regularly check the internal temperature and turn over the mixture when it reaches 140°F. The compost pile should be built in layers 3 - 4 inches deep. Composting still happens if the pile is not turned, but the materials break down slowly.

Step 5: Using compost around your home.

Once the composting process is complete, the result is a dark, nutrient-rich humus that has many uses:

- Soil Amendment: Work a 1 - 3 inch layer of compost into garden soil.
- Mulch: Apply a 2 - 3 inch layer on top of existing soil.
- Potting Mix: Blend with potting soil for container plants.

Waste Minimization

Traditionally, waste is viewed as an unnecessary element arising from the activities of any industry. In reality, waste is a misplaced resource, existing at a wrong place at a wrong time.

Waste is also the inefficient use of utilities such as electricity, water, and fuel, which are often considered unavoidable overheads. The costs of these wastes are generally underestimated by managers. It is important to realise that the cost of waste is not only the cost of waste disposal, but also other costs such as:

- Disposal cost.
- Inefficient energy use cost.
- Purchase cost of wasted raw material.
- Production cost for the waste material.
- Management time spent on waste material.
- Lost revenue for what could have been a product instead of waste.
- Potential liabilities due to waste.

Waste minimisation can be defined as "systematically reducing waste at source". It means:

- Prevention and reduction of waste generated.
- Efficient use of raw materials and packaging.
- Efficient use of fuel, electricity and water.
- Improving the quality of waste generated to facilitate recycling and reduce hazard.
- Encouraging re-use, recycling and recovery.

Waste minimisation is also known by other terms such as waste reduction, pollution prevention, source reduction and cleaner technology. It makes use of managerial and technical interventions to make industrial operations inherently pollution free.

It should be also clearly understood that waste minimization, however attractive, is not a panacea for all environmental problems and may have to be supported by conventional treatment/disposal solutions.

Waste minimization is best practiced by reducing the generation of waste at the source itself. After exhausting the source reduction opportunities, attempts should be made to recycle the waste within the unit. Finally, modification or reformulation of products so as to manufacture it with least waste generation should be considered. Few wastes and possible resources are indicated in the table.

Table: Waste and possible resources.

Wastes	Resources
Fly ash from power plant.	Raw material for cement or brick manufacture.
Bagasse wastes from sugar manufacture.	Fuel for boiler.
CO_2 release from ammonia plant.	Raw material for Urea manufacture.

Classification of Waste Minimization (WM) Techniques

The waste minimization is based on different techniques. These techniques are classified as hereunder.

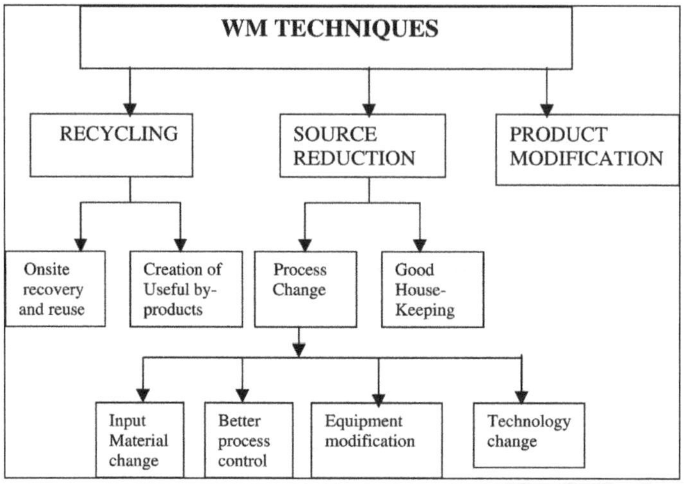

Source Reduction

Under this category, four techniques of WM are briefly discussed below:

- Good Housekeeping: Systems to prevent leakages and spillages through preventive maintenance schedules and routine equipment inspections. Also, well-written working instructions, supervision, awareness and regular training of workforce would facilitate good housekeeping.
- Process Change: Under this head, four CP techniques are covered:
    - Input Material Change: Substitution of input materials by eco-friendly (nontoxic or less toxic than existing and renewable) material preferably having longer service time.
    - Better Process Control: Modifications of the working procedures, machine-operating instructions and process record keeping in order to run the processes at higher efficiency and with lower waste generation and emissions.
    - Equipment Modification: Modification of existing production equipment and utilities, for instance, by the addition of measuring and controlling devices, in order to run the processes at higher efficiency and lower waste and emission generation rates.
    - Technology Change: Replacement of the technology, processing sequence and synthesis route, in order to minimise waste and emission generation during production.
- Recycling:
    - On-site Recovery and Reuse: Reuse of wasted materials in the same process or for another useful application within the industry.
    - Production of Useful by-product: Modification of the waste generation process in order to transform the wasted material into a material that can be reused or recycled for another application within or outside the company.
    - Product Modification.

Characteristics of the product can be modified to minimise the environmental impacts of its production or those of the product itself during or after its use (disposal).

Waste Minimization Methodology

For an effective waste minimization programme, it is essential to bring together various groups in the industry to ensure implementation. How formalised the programme would be depends upon the size and composition of the industry and its waste and emission problems. The programme should be flexible enough so that it can adapt itself to changing circumstances. A methodical step-by-step procedure ensures exploitation of maximum waste minimization opportunities. The steps in a typical waste minimization progamme are illustrated below:

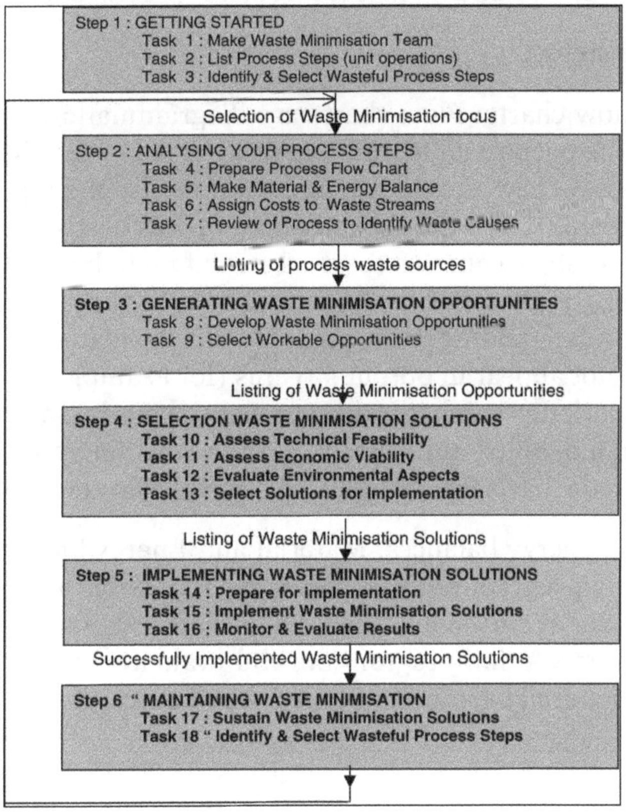

Step 1: Getting started.

- Form a waste minimization team: Waste Minimization (WM) concept encompasses all departments and sections in an industry. Activities involved in WM audits require assistance and cooperation not only from the personnel belonging to concerned department, but also from other related departments. Hence, making an inter-disciplinary and inter-departmental team is a prerequisite for successful conduct of a WM audit. In special cases, it would be advantageous to have external experts in order to have an unbiased, professional approach.
- List process steps/unit operations: The WM team should familiarize itself with the manufacturing processes including utilities, waste treatment and disposal facilities, and list all the process steps. Preparation of sketches of process layout drainage system, vents and material-loss areas would be useful. This helps in establishing cause-effect relationships and

ensuring that important areas are not overlooked. Periodic, intermittent and continuous discharge streams should be appropriately labeled.

- Identify and select wasteful process steps: In multi-process type industries, it may be difficult to start detailed Waste minimization activities covering the complete unit. In such cases, it is advisable to start with fewer process steps to begin with. The selected step(s) could be the most wasteful and one with very high waste minimization potential.

This activity could also be considered a preliminary prioritization activity. All the various wasteful steps identified in 1.2 should be broadly assessed in terms of volume of waste, severity of impact on the environment, Waste minimization opportunities, estimated benefits (specially cost savings), cost of implementation etc. Such assessment would help in focusing on the process steps/areas for detailed analysis.

Step 2: Analysing process steps.

- Prepare process flow charts: Flow charts are diagrammatic/schematic representation of production, with the purpose of identifying process steps and the source of waste streams and emissions. A flow chart should list, and characterize the input and output streams, as well as recycle streams. Even the so called free or less costly inputs like water, air, sand, etc should be taken into account as these often end up in being the major cause of wastes. Wherever required, the process flow diagram should be supplemented with chemical equations to facilitate understanding of the process. Also the materials which are used occasionally and which do not appear in output streams (for example catalysts, coolant oil) should be specified. The periodic/batch/continuous steps should also be appropriately highlighted. Preparation of a detailed and correct process flow diagram is a key step in the entire analysis and forms the basis for compilation of materials and energy balance.

- Make material and energy balances: Material and Energy balances are important for any Waste minimization programme since they make it possible to identify and quantify, previously unknown losses or emissions. These balances are also useful for monitoring the progress achieved in a prevention programme, and evaluating the costs and benefits. Typical components of a material balance and energy balance are given below:

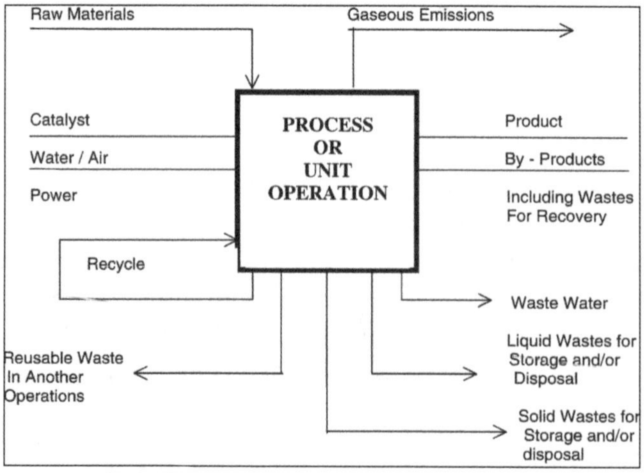

Typical components of a material balance.

It is not advisable to spent more time and money to make a perfect material balance. Even a rough/

preliminary material balance throws open Waste Minimization opportunities which can be profitably exploited.

On the other hand, the precision of analytical data and flow measurements is important as it is not possible to obtain a reliable estimate of the waste stream by subtracting the materials in the product from those in the raw materials. In such cases, a direct monitoring and analysis of waste streams should be carried out.

- Assign costs to waste stream: In order to assess the profit potential of waste streams, a basic requirement would be to assign costs to them. This cost essentially reflects the monetary loss arising from waste. Apparently, a waste stream does not appear to have any quantifiable cost attached to it, except where direct raw material/product loss is associated with it. However, a deeper analysis would show several direct and indirect cost components associated with waste streams such as:
- Cost of raw materials in waste.
- Manufacturing cost of material in waste.
- Cost of product in waste Cost of treatment of waste to comply with regulatory requirements.
- Cost of waste disposal.
- Cost of waste transportation.
- Cost of maintaining required work environment.
- Cost due to waste cess.

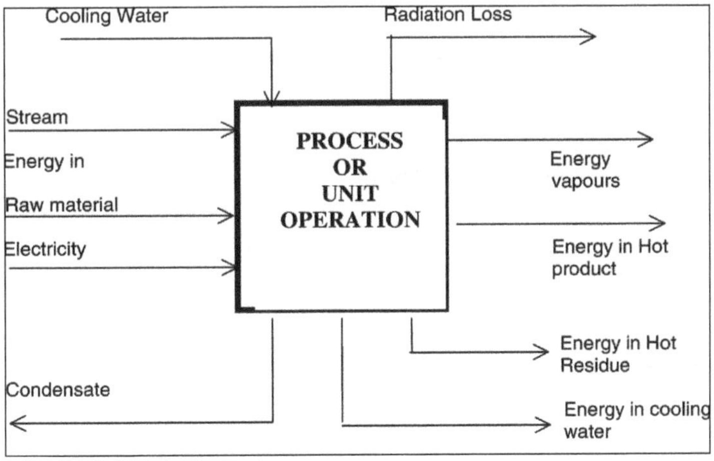

Typical components of an energy balance.

Based on this, for each waste stream, total cost per unit of waste (Rs/Kl or Rs/Kg) should be worked out. This figure would be useful in working out the feasibility of the waste minimization measures. The result can also be used to categorize the waste streams for priority action.

- Review of process: Through the material and energy balances, it is possible to carry out a cause analysis to locate and pinpoint the causes of waste generation. These causes would subsequently become the tools for evolving Waste Minisation measures. There could be a wide variety of causes for waste generation ranging from simple lapses of housekeeping to complex technological reasons.

Typical Causes of Waste

Poor housekeeping:

- Leaking taps/valves/flanges.
- Spillages.
- Overflowing tanks.
- Worn out material transfer belts.

Operational and maintenance negligence:

- Unchecked water/air consumption.
- Unnecessary running of equipment.
- Sub optimal loading.
- Lack of preventive maintenance.
- Sub-optimal maintenance of process condition.
- Ritualistic operation.

Poor raw material quality:

- Use of substandard cheap raw material.
- Lack of quality specification.
- Improper purchase management system.
- Improper storage.

Poor process/equipment design:

- Mismatched capacity of equipment.
- Wrong material selection.
- Maintenance prone design.
- Adoption of avoidable process steps.
- Lack of information/design capability.

Poor layout:

- Unplanned/adhoc expansion.

- Poor space utilization plan.
- Bad material movement plan.

Bad technology:

- Continuation of obsolete technology.
- Despite product/raw material change.
- High cost of better technology.
- Lack of availability of trained manpower.
- Small plant size.
- Lack of information.

Inadequately trained personnel:

- Increased dependence on casual/contract labour.
- Lack of formalized training system.
- Lack of training facilities.
- Job insecurity.
- Fear of losing trade secrets.
- Lack of availability of personnel.
- Understaffing hence work over pressure.

Employee demotivation:

- Lack of recognition.
- Absence of reward.
- Emphasis only on production, not on people.
- Lack of commitment and attention by top management.

Step 3: Generating waste minimization opportunities.

- Develop waste minimization opportunities: Once the origin and causes of waste and emissions are known, the process enters the creative phase. The WM team should now start looking for possible opportunities for reducing waste. Finding potential options depends on the knowledge and creativity of your team members. The potential sources of help in finding Waste Minimization Opportunities are:
    - Other personnel from the same or similar plant elsewhere.

- ◦ Trade associations.
- ◦ Consultancy organizations.
- ◦ Equipment suppliers.
- ◦ Consultants.

The process of finding Waste Minimization opportunities should take place in an environment, which stimulates creativity and independent thinking. It would be beneficial to move away from the routine working environment for better results. Various analysis tools and techniques like "brainstorming", "group discussions" etc would be useful in this step.

- Select workable opportunities: The Waste Minimization opportunities developed should be screened and those, which are impractical, should be discarded.

The discarding process should be simple, and straightforward and may often be only qualitative. There should be no ambiguity or bias. Unnecessary effort in doing detailed feasibility analysis of opportunities, which are impractical or non-feasible, should be avoided. The remaining Waste Minimization opportunities are then subjected to a more detailed feasibility analysis.

Step 4: Selecting waste minimization solutions.

The selection of a Waste Minimization solution for implementation requires that it should not only be techno-economically viable, but also environmentally desirable.

- Assess technical feasibility: The technical evaluation determines whether a proposed Waste Minimization option will work for the specific application. The impact of the proposed measure on product production rate should be evaluated first. In case of significant deviation from the present process practices, laboratory testing trial runs might be required to assess the technical feasibility.

  A typical checklist for technical evaluation could be as follows:

  - ◦ Availability of equipment.
  - ◦ Availability of operating skills.
  - ◦ Space availability.
  - ◦ Effect on production.
  - ◦ Effect on product quality.
  - ◦ Safety aspects.
  - ◦ Maintenance requirements.
  - ◦ Effect on operational flexibility.
  - ◦ Shut down requirements for implementation.

- Assess economic viability: Economic viability would often be the key parameter to promote or discourage Waste Minimization. For a smooth take-off, it is therefore essential that the first few Waste Minimization measures should be economically very attractive. Such a strategy helps in creating more interest and sustains commitment.

Options requiring little investment, but involving more of procedural changes (housekeeping measures, measures, operational improvements) do not need an intensive economic analysis and simple methods like pay back period could be used. However as Waste Minimization measures become more involved and capital intensive, other profitability analysis methods viz. Return on Investment (IRR) or Net Present Value (NPV) are recommended to get the total picture.

While doing the economic investment, the costs may include fixed capital i.e. cost of equipment, working capital cost, shutdown cost, O and M costs etc. The savings could be direct savings of input materials/energy, increased production levels and hence lower specific input cost, lower O and M cost, value of by products, reduction in environmental cost i.e. waste treatment transportation and disposal cost etc.

- Evaluate environmental aspects: The options for Waste Minimization with respect to their impact on the environment should be assessed. In many cases the environmental advantage will be obvious if there is a net reduction in the toxicity and quantity of waste. Other impacts could be changes in treatment of the wastes. In the initial stages, environmental aspects may not appear to be as compelling as economic aspects. In future as in developed countries, environmental aspects would become the most important criteria irrespective of the economic viability.
- Select Solutions for Implementation: After technical, economic and environmental assessment, Waste Minimization measures should be selected for implementation. Understandably, the most attractive ones would be those having best financial benefits, provided technical feasibility is favourable. However, in a growing number of cases especially when active pressure groups are present, environmental priorities may become the final deciding factor.

The work done so far should be documented. Apart from becoming a reference document for seeking approvals in implementation, the document would also be useful in obtaining finances from external finance institutions, reporting status to other agencies, and establishing base levels for performance evaluation and review.

Step 5: Implementing waste minimization solutions.

- Prepare for implementation: The selected solutions could be taken for implementation. Apart from simple housekeeping measures several others would require a systematic plan of implementation.

The Waste Minimization team should be well prepared to take up the job of implementation. The preparation would include arranging finances, establishing linkages in case of multidepartment solutions, technical preparations, etc.

- Implement solutions: The task comprises layout and drawing preparation equipment fabrication/procurement, transportation to site, installation and commissioning. Whenever required, simultaneous training of manpower should be taken up as many excellent measures have failed miserably because of non-availability of adequately trained people.
- Monitor and evaluate results: The WM solutions should be monitored for performance.

The results obtained should be matched with those estimated/worked out during technical evaluation to establish causes for deviation, if any. The implementation job is considered to be over, only after successful commissioning and sustained stable performance over a reasonable length of time.

Step 6: Sustain waste minimization.

The biggest challenge in Waste Minimization lies in sustaining Waste Minimization. The enthusiasm of the Waste Minimization team wanes off with time. Such tragic ends should be avoided. Backing out from commitment, predominance of production at any cost, absence of rewards and appreciation, and shifting of priorities are some of the commonly encountered reasons, which one should check and avoid.

Also monitoring and review of the implemented measures should be communicated to all employees in the industry so that it fans the desires for minimizing wastes. Involvement of as large a number of employees as possible and rewarding the deserving ones, will help long term sustenance of Waste Minimization.

Having implemented Waste Minimization solutions in the area under study, the Waste Minimization team should go back to Step-2 i.e. analysing the process steps and identifying and selecting the next wasteful area. In this way, the cycle continues, till all the steps are exhausted.

In a nutshell, a philosophy of minimizing waste must be developed within the company. This means that Waste Minimization should become an integral part of company's activities.

MINING WASTEWATER TREATMENT

The mining industry has contributed tremendously to the global economy. However, waste generated by this activity poses many challenges. Current low-cost technologies for the removal of heavy metals from mining wastewater include biosorption, adsorption, CWs (Constructed Wetlands) and waste stabilization ponds.

Heavy metals occur in immobilized form in sediments and as ores in nature. Various human activities like ore mining and industrial processes disrupt the natural biogeochemical cycles causing increased deposition of heavy metals in terrestrial and aquatic environments. Heavy metals are found in wastewater discharged from industries such as mining, metal processing, electroplating, textile, tannery and petroleum refinery. Heavy metals such as As, Cu, Pd, Zn, Fe, Cr, Ni and Hg are toxic and therefore the discharge of untreated metal-laden effluent from the mining industry poses serious environmental challenges. Sorption using natural and agricultural materials has been proposed as a cost effective and environmentally friendly alternative to chemical treatment.

The discovery and further development of biosorption phenomena provide a basis for a whole new technology aimed at the removal of various pollutants or the recovery of valuable resources from aqueous systems. Although there are many industrial processes using sorbents such as activated carbon, the development of biosorption processes is still mainly at the stage of laboratory studies in spite of enormous progress made over the last decade. Most studies on continuous biosorption systems have focused mainly on the optimisation of operating conditions and breakthrough curves using synthetic metal solutions. Many of these studies have limited industrial application because industrial effluents are more complex, containing several metal ions and other contaminants. Treatment of industrial effluent in continuous flow through systems allows the use of biosorbents for multiple biosorption and desorption applications. In this way, process sustainability is achieved through a substantial reduction in the fresh sorbent requirement, operational cost and solid waste materials needing disposal or containment.

WSPs (Waste Stabilization Ponds) are common technologies for industrial wastewater treatment due to their low capital and operating costs. They are particularly popular in locations where large tracts of land are available and affordable. In spite of the popularity of WSPs, limited studies on the use of this technology for the removal of heavy metals from mining wastewater had been conducted. The few studies conducted on WSPs have shown a poor performance in the removal of metals. Due to this limitation of scanty literature availability, greater attention is paid to the use of CWs (Constructed Wetlands) for mining wastewater treatment.

CWs are artificial or engineered wetlands developed to utilize the natural, physical, biological and chemical processes that take place in these aquatic ecosystems involving the soil, flora and fauna, including its microbial communities, to treat wastewater. The use of CWs for the removal of heavy metals is a very recent development. CWs are used for the treatment of wastewater from mining, oil refineries, chemical factories, pulp and paper production, tannery, textile, abattoir, distillery, winery and other industries as well as various runoff waters from airports and highways as various sources of industrial wastewater, noting limited studies done on wastewater from mines. This is because the treatment of wastewater from mines had not been given much attention, as results that monitoring programmes by mining companies are usually not published in peer reviewed journals.

Technological Options for the Removal of Heavy Metal from Mining Wastewater

Currently, sustainable technologies for the removal of heavy metals from mining wastewater include biosorption, adsorption, CWs and waste stabilization ponds, among others.

Biosorption

Wang, J. and Chen, C. defined biosorption as the removal of metal and metalloid species, compounds and particulates from solution by biological material. Biosorption is made possible by the ability of biological materials to accumulate heavy metals from wastewater through metabolically mediated or physico-chemical uptake pathways.

Understanding Metals Biosorption Mechanism during Mining Wastewater Treatment due to the interaction of several factors on specific biosorbents, it is almost impossible to propose a general mechanism. Although several metal-binding mechanisms have been put forward, e.g. physical and chemical adsorption, ion exchange and microprecipitation, the actual mechanism of metal biosorption is still not fully understood.

Ion-exchange is an important concept in biosorption, because it explains many observations made during heavy metal uptake experiments. Ion-exchange is a reversible chemical reaction where an ion within a solution is replaced by a similarly charged ion attached onto an immobile solid particle. This ion-exchange process is one of the well-known surface reactions, which represents a great degree of complexity, primarily due to its multi-species nature. In bacterial biosorption, the bacterial cell wall is the first component that comes into contact with the metal, where the solute can be deposited on the surface or within the cell wall structure. Since the mode of solute uptake by dead or inactive cells is extracellular, the chemical functional groups of the cell wall play vital a role in biosorption.

Another frequently encountered metal binding mechanism is chelation, which can be defined as a firm binding of metal ions with an organic molecule (ligand) to form a ring structure. Various functional groups including carboxylic, hydroxyl, sulphate, phosphate, amides and amino groups can be considered possible for sorption. Among these groups, the amino group is the most effective for removing heavy metals, since it does not merely chelate cationic metal ions but also absorbs anionic species through the electrostatic interaction or hydrogen bonding.

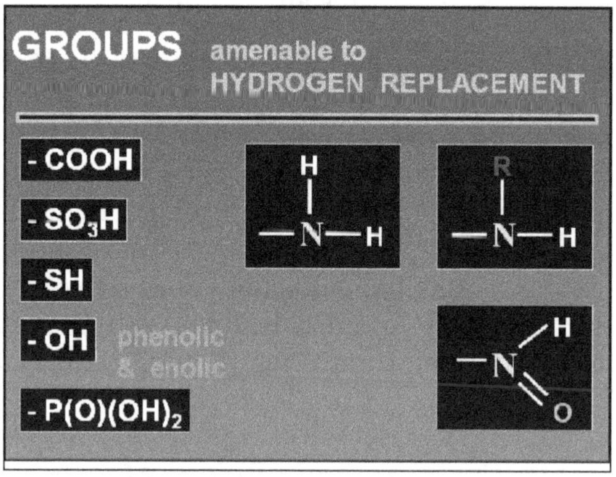

Major binding groups for biosorption.

As they are negatively charged and abundantly available, carboxyl groups actively participate in the binding of metal cations. Figure shows major functional groups involve in metal binding.

Agricultural and plant waste materials, such as coconut husk and shell, are usually composed of lignin and cellulose as the main constituents. Other components present in the metal binding process are hemicellulose, lipids, proteins, simple sugars, starches, water, hydrocarbons and many more compounds that contain a variety of functional groups. The presence of these functional groups and their complexation with heavy metals during biosorption has been reported by different researchers using spectroscopic techniques.

Benavente, M., reported the treatment of gold mining wastewater using chitosan and found that the removal of Cu, Pb and Zn was not affected by the presence of other metals because they were present at low concentrations. The speciation of metallic ions in gold mining tailings wastewater depends on their oxidation states, the metallic ion concentrations, the cyanide concentration, the redox potential, the temperature and pH of the wastewater. Benavente, M., indicated that the presence of cyanide had a negative impact on the sorption process, but not to the extent of affecting the removal of copper significantly, as evident from the low copper concentration (< 0.01 mg·L-1) in their treated effluent. Similar to the findings of Benavente, M.

Acheampong, M. A. and Lens, P. N. observed that neither the presence of cyanide nor other heavy metals inhibited the removal of copper from gold mining effluent studied.

According to Saad, D. M., metal-cyanide complexes and free metallic ions were mainly present in the gold mining wastewater at pH 7.7. They indicated that the speciation results for the tailings dam show that copper, lead and zinc were mostly found as free metallic ions with only a small fraction present as metal-cyanide complexes. This is, however, contradictory to well-established experimental science of metallocyanide species in tailings waters. The most commonly encountered iron cyanide complexes are the two formed with ferrous and ferric iron, $Fe(CN)_6^{4-}$ and $Fe(CN)_6^{3-}$, respectively. The Fe(III) complex being the predominant species is found in mining tailings solutions. The formation of hexacyanoferrate(III) is primarily the result of the oxidation of hexacyanoferrate(II) in surface impoundments. However, typically within solid tailings environment, the oxidation potential would result in the formation of ferrocyanide precipitates. The hexacyanoferrates are classified as "inert" complexes, in that their chemical stability results from extremely slow rates of dissociation and relatively low toxicity. The overall effect is the reduction in the free metallic ions present and an increase in the proportion of the metal-cyanide complexes in the tailings solutions.

When present in cyanide tailings solutions, copper is predominantly present as a mix of three forms: $Cu(CN)^{2-}$, $Cu(CN)_3^{2-}$ and $Cu(CN)_4^{3-}$. The Cu(II) cyanide species are unstable in aqueous solution and disproportionate to the Cu(I) cyanides. While the formation of metal cyanide complexes does not completely eliminate the toxicity of cyanide, it does effect a substantial reduction in such toxicity. Metal cyanide complexes are also intermediates involved in the formation of more stable compounds that remove free cyanide from the tailings environment. Kunz, R. G., described a process where copper was complexed with cyanide to enhance the adsorption and oxidation of cyanide onto activated carbon. The environmental significance of complexation of cyanide with heavy metals is that it is relatively rapid and it occurs wherever soluble species of these metals are present, such as found in gold mining tailings. It reduces the toxicity of free cyanide and acts as an intermediary to reduce cyanide mobility via adsorption onto organic and inorganic surfaces or precipitation of metal cyanide salts.

In treating electroplating wastewater using sugar cane bagasse, Sousa, F. W., reported a considerable decrease in adsorption of zinc and nickel, probably due to electrostatic repulsive forces between the cationic sugar cane bagasse surface and the metal ions. In contrast, Acheampong, M. A. and Lens, P. N. reported that coconut shell used in their study was negatively charged at the gold mining effluent pH due to the lower point of zero charge pH of 6.5. Ahmed, M. J. K. and Ahmaruzzaman, M. suggested that the removal of the positively charged heavy metals from gold mining effluent could be enhanced by electrostatic attraction forces between the biosorbent and the sorbate.

Process Conditions

The investigation of factors affecting the efficiency of heavy metal sorption is of great interest for the industrial community. The efficiency is strongly influenced by the physico-chemical characteristics of the solutions, such as pH, temperature, initial metal concentration, presence of other ions and sorbent dosage. These factors are important in evaluating the maximal sorption performance of any sorbent. The solution pH is a crucial factor in heavy metal sorption. The pH value significantly influences the dissociation of the sorbent and the solution chemistry of the heavy metals, i.e. metal speciation, hydrolysis, complexation by organic and inorganic ligands, redox reactions and precipitation are all pH dependent. Competition between cations and protons for binding sites means that sorption of metals like Cu, Cd, Ni, Co and Zn is often reduced at low pH values. Due to the importance of pH in sorption, many researches were conducted on its effect on the removal of various heavy metals.

Depending on the structure and surface functional groups of a sorbent, temperature has an impact on the adsorption capacity within the range of 20-35 o C. It is well known that a temperature change alters the adsorption equilibrium in a specific way determined by the exothermic or endothermic nature of a process. Higher temperatures usually enhance sorption due to the increased surface activities and kinetic energy of the solute. However, physical damage can be expected at higher temperatures. It is always desirable to evaluate the sorption performance at room temperature, as this condition is easy to replicate.

Another important parameter in biosorption is the ionic strength, which influences the adsorption of solute to the biomass surface. Industrial wastewater often contains ions other than heavy metal ions, e.g., Na^+, K^+, Mg^{2+} and Ca^{2+}, which may interfere with heavy metal ion uptake by biomass. As a general trend, the metal uptake is found to decrease with increasing ionic strength of the aqueous solution as a result of more electrostatic attraction and change of the metal activity.

The amount of biomass in the solution also affects the specific metal uptake. For lower values of biomass concentrations, there is an increase in the specific uptake. Gadd, G. M., suggested that an increase in biomass concentration leads to interference between the binding sites. By increasing the adsorbent dosage, the adsorption efficiency increases even though the amount adsorbed per unit mass decreases. In principle, with more adsorbent present, the available adsorption sites or functional groups also increase. In turn, the amount of adsorbed heavy metal ions increases, which results in an improved adsorption efficiency.

The size of the biosorbent also plays a vital role in biosorption. Smaller sized particles have a higher surface area, which in turns favours biosorption and results in a shorter equilibrium time. Simultaneously, a particle for biosorption should be sufficiently resilient to withstand the application of pressure and extreme conditions applied during regeneration cycles. Therefore, preliminary experiments are necessary to determine the suitable size of a biosorbent.

Process Con iguration

Due to the concentration difference driving force for sorption (between the sorbate in the solution and that already sequestered), generally the most effective configuration of the sorption system is that based on a flow-through fixed bed type of a reactor/contactor. The sorption column would generally not exceed approximately 1.8 m in diameter and 4-5 m in height. The scaling up of the process to accommodate larger flows is attained by simply multiplying the number of columns that would operate in parallel. Rarely, usually because of special process requirements, a mixed contactor could also be used. Different, usually multi-stage, counter-current configurations could then be arranged. The most important criterion in the design of fixed-bed sorption systems is the prediction of the column breakthrough or the shape of the adsorption wave front, which determines the operating life-span of the bed. Figure shows a typical of fixed-bed sorption systems for treating gold mining effluent.

Process Performance

The overall performance of the processes based on the sorption column mode of operation is judged by the column operating time (until the column sorption capacity is exhausted), characterized by the sorbate break-through point as determined at the column exit. Since the fixed-bed column does get eventually saturated, a process arrangement is often used whereby one column is in active sorption operation while another one is being prepared (regenerated and cleaned) during its stand-by period. That column preparation usually consists of elution of the sequestered sorbate with simultaneous or sequential sorbent regeneration, clean-up and washing. Sorbent regeneration capacity is an important characteristic that results in a better overall process economy.

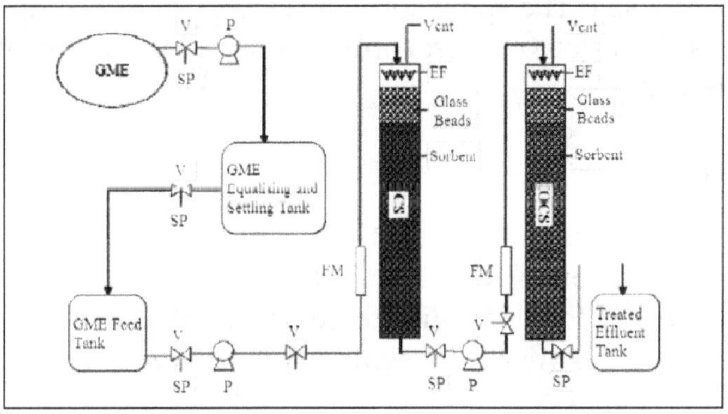

Schematics diagram of a two-stage pilot plant set-up for treating gold mining effluent (CS: Coconut Shell; IOCS: Iron Oxide Coated Sand; EF: Effluent Distributor; P: Pump; SP: Sampling Point; V: Valve; GME: Gold Mining Effluent; FM: Flow Meter).

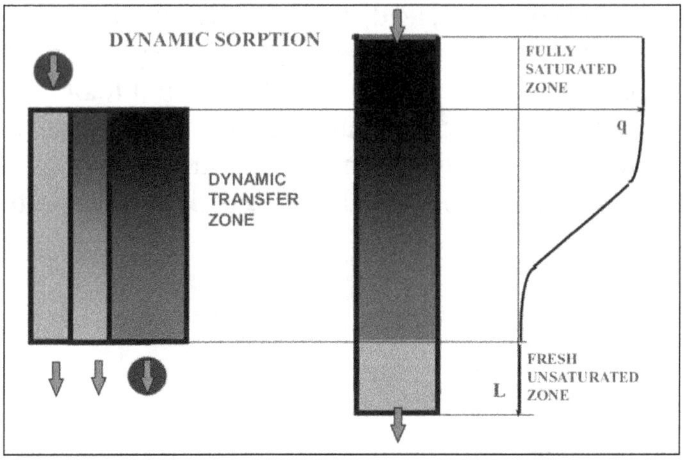

Schematic representation of the development of dynamic transfer zone in the continuous-flow fixed bed sorption column.

Sorbent Regeneration and Metal Recovery

One of the main attributes of biosorption is the potential of sorbent regeneration and metal recovery. Desorption of sorbed metals from loaded sorbent enables reuse of the sorbent and recovery and containment of the sorbed metals from a highly concentrated eluant stream. In some cases, desorption treatment may further improve the sorption capacities, although in other cases there may be a loss of sorption efficiency and even a complete disintegration of the biosorbent. To improve process viability, better understanding of the sorption-desorption processes in sorbent regeneration is needed. The acid-based metal desorption may be followed by further steps to prepare the biosorbent for the most efficient subsequent uptake cycle. This may entail pre-saturating the active sites of the biomaterial with more pH-neutral calcium or potassium cations in order to avoid an uncontrollable acidic wave inside the column caused by the release of ion-exchanged protons. Some overall process optimization may be useful in this respect.

Metal recovery is a down-stream consideration that has been added to the metal removal aspect, which is driven predominantly by environmental detoxification considerations. The sorption and desorption processes offered the opportunity for an economical recovery of metals from the concentrated solution. The feasibility of the recovery operation depends, to a large degree, on the type of recovery procedures used. The spent sorbents generated at the end of the process must be properly and safely disposed. Various disposal options that are in conformity with local solid waste management practice are employed.

Process Integration

For a successful industrial application biosorption technology for the sustainable treatment of mining wastewater, the integration of the sorbent processing and formulation, sorption-desorption application, metal recovery for reuse and the disposal of the spent biomass is very critical. Where the treated effluent does not meet the regulatory discharge requirement, a further polishing step using WSPs and CWs is applied.

WSPs and CWs

The extraction of metals such as gold, zinc, lead, copper, iron, aluminium and many others involves the release of many contaminants that may pose serious threat to public health or wildlife and therefore had to be removed or converted to various forms that are not toxic. The contaminants include heavy metals such as zinc, copper, lead, chromium, iron and manganese, cyanide, phosphate, carbonates, sulphides, sulphates, arsenic and its complexes, nitrogen and its compounds. A detail list of the concentrations of the constituents of typical mining activity wastewater is outlined in table.

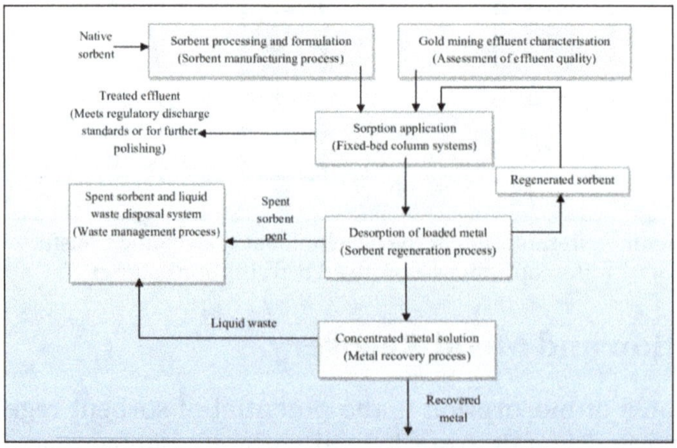

Generalised integrated single stage sorption systems for gold mining effluent treatment using locally available low-cost materials.

WSPs are common technologies for wastewater treatment due to their low capital and operating costs. They are particularly popular in locations where large tracts of land are available and affordable. Inspite of the popularity of WSPs, very limited studies on the use of this technology for the removal of heavy metals as pertains in mining wastewater had been conducted. The few studies conducted on WSPs have shown a poor performance in the removal of metals. Due to this limitation of scanty literature availability, greater attention would be paid to the use of CWs.

CWs are artificial or engineered wetlands developed to utilize the natural, physical, biological and chemical processes that take place in these aquatic ecosystems involving the soil, flora and fauna, including its microbial communities to treat wastewater. The use of CWs for the removal of heavy metals is a very recent development. In his review of the use of CWs in industrial wastewater treatment, Vymazal, J. mentioned the oil refineries, chemical factories, pulp and paper production, tannery and textile industries, abattoir, distillery, winery and other food processing industries as well as various runoff waters from airports and highways as various sources of industrial wastewater, noting limited studies done on wastewater from mines. This is because the treatment of wastewater from mines had not been given much attention in literature, as results that monitoring programmes by mining companies are usually not published in peer reviewed journals. This notwithstanding, a few studies had been documented. For example, Pantano, J., were successful in using a horizontal flow CW to remove Cd, Zn and Cu from mining impacted groundwater while Gerth, A., used a hybrid free water surface and horizontal flow CW to treat seepage water from a uranium mine in Germany.

Table: Typical constituents of wastewater from different mining activity.

Key parameters	Sources of wastewater		
	Gold mining	Zinc and lead mining	Zinc-rich acid mine drainage
pH	701-8.6	8.23	3.58
Conductivity	94-979 mS/m		1,915 µS/cm
TSS		2,495 ± 182 mg/L	
Mg	31-1,488 mg/L		19.0 mg/L
Ca			136 mg/L
Li	0.22-2.8 mg/L		
Na	14-114 mg/L		14.3 mg/L
K	4.7-114 mg/L		12.3 mg/L
Al			12.2 mg/L
As	41-40 µg/L	0.077 ± 0.024 mg/L	
Sb	35-160 mg/L		
Zn		14.47 ± 8.37 mg/L	552 mg/L
Pb		11.49 ± 7.7 mg/L	
Hg		0.0003 ± 0.0002 mg/L	
Ni	37-69 µg/L		
Cd		0.05 ± 0.04 mg/L	
Fe	67-109 µg/L		
Mn	0.42-1.3 mg/L		17 mg/L

Treatment Performance and Limitations

The removal of toxic metals from mining wastewater using physico-chemical approaches such as adsorption, oxidation and reduction and chemical precipitation require large quantities of chemical reagents and produce a considerable amount of toxic sludge and secondary pollutants, thus raising questions about the sustainability of these technologies. Conventional wastewater treatment systems such as the activated sludge treatment system are known to perform better in the removal of heavy metals than WSPs.

One limitation of WSPs in heavy metal removal is its inability to overcome shock loadings due to the low presence of microorganisms. This limitation was overcomed by the introduction of attached growth medium which resulted in significant increase in heavy metal removal. In the absence of such modifications, WSPs had to operate at very low organic loadings for metal removal. The use of CW in treating mining wastewater is also limited to the polishing stages of treatment as influent of high suspended solids can lead to clogging.

Ustun, G. E. reported in a study involving nine metals (Al, Cd, Cr, Cu, Fe, Mn, Ni, Pb and Zn) that while the WSP removed over 50% in Al and Cr, for the other metals removals were less than 20% compared with the activated sludge system which a range of removals from 47% in Ni to 95% for Cr. Polprasert, C. and Charnpratheep, K. observed that when a medium is provided in WSPs for the adsorption of metals, the performance of the WSPs can increase significantly. Sekomo, C. B., using synthetic wastewater, however, reported a removal rate of Cr of 94% and 98% for a duckweed and algal pond systems respectively. The researchers noted that the rates of removal obtained occurred irrespective of the loading rate or the light regime. In that same study, using a light-darkness regime of 16 h light/8 h darkness, Sekomo, C. B., recorded a Zn removal rate of 70% at a lower loading rate, which dropped to a removal rate of 40% when the loading rate was increased. The removal efficiency rose back to 80% at the higher metal loading rate but under 24 h light regime. Pb, Cd and Cu all showed relatively similar patterns with removal efficiencies of 36% and 33% for Pb, 33% and 21% for Cd and 27% and 29% for Cu in the duckweed and the algal ponds, respectively in that study. In biological wastewater treatment, metal removal efficiency depends on the metal species and concentration, the reactivity of the available biopolymers or biomass and the composition of other wastewater components. Sekomo, C. B., however, observed that inspite of the physico-chemical differences among the duckweed and algal ponds, differences in metal removal efficiency between algal and duckweed ponds were rather small. CW offers a viable alternative for the removal of metals from mining wastewater.

Mechanisms of Metal Removal

In waste stabilization ponds, the removals of metals occur through a process of adsorption unto suspended solids and biomass, with subsequent sedimentation of adsorbed metals. This takes place mainly in the anaerobic pond and to some extent the facultative ponds. The metal removal efficiency depends on the metal species and concentration, the reactivity of the available biopolymers or biomass and the composition of other wastewater components. Thus the formation of metal complexes and dissolution values for each metal are different resulting in varying removal efficiency for each metal type. As mentioned earlier, the low amount of medium for adsorption by metals in WSPs limits the extent of its removal in WSPs.

Mechanisms involved in heavy metal removal in CWs, include a combination of sedimentation, filtration, binding to substrata, precipitation as insoluble salts such as sulphides or oxyhydroxides and accumulation by plants, algae and bacteria.

It has been argued that substrates act as primary sink for heavy metal accumulation and that the bulk of the influent heavy metal is removed by accumulation in the ground biomass and not the above ground biomass of plants.

Heavy metal uptake varies from one plant species to another in CWs. However, Yadav, A. K., noted that in all three plant species involving Canna indica, Typha angustifolia and Cyperus alternifolius in a VSSF (Vertical Subsurface Flow), the contribution of the below ground biomass (roots) was more than the above ground biomass (leaves and stem) in the removal of Cu, Cr, Co, Ni and Zn from aqueous solution. They also noted that the removal of Cr, Ni, Cu, Zn and Co was enhanced by 16.6%, 22.9%, 20.4%, 21.5% and 21.8% respectively, when gravel bed depth of CWs was increased from 0.3 m to 1.5 m, showing that enhanced gravel bed depth may improve the performance of CWs in metal removal.

In addition to taking up metal, the root network also filters fine particulate matter, for which sedimentation is slower than for larger particles. Through the biofilm that forms on the surface of the roots, significant biological activity develops and could intervene in the process of biodegradation of other pollutants, such as hydrocarbons.

It has been suggested that plants used for the uptake of heavy metals must be mowed yearly to harvest biomass and recover pollutants and these plants used for providing fuel.

Economics of Heavy Metal Removal

In determining the suitability of a technology for treating wastewater, key considerations are given to its robustness while meeting effluent standards, extent of chemical usage and the degree of environmental nuisance it poses, capacity for various re-use options as well as environmental benefits, greenhouse gas emissions and its energy and capital requirements. The high purification rates, robustness and operational stability, low skilled manpower and capital requirement, ecological benefits as well as direct economic benefits such as fish farming of WSPs and CWs are well documented in literature. For the treatment of mining wastewater, however, particular importance is attached to the energy requirement as well as the carbon footprint of the treatment technology.

Table: Performance of various types of CWs in the removal of metals from mining wastewater.

Metal/CW type	Influent csoncentration (mg/L)	Retention time	Removal (%)
Al/VSSF	0.51	79 days	100.0
As/HSSF	2.3	11 days	99.9
Zn/VSSF	3.19	36 days	100.0
Zn/VSSF	20.0	72 hours	99.3
Zn/HSSF	14.5	5-7 days	97.3
Pb/VSSF	0.0105	36 days	100.0
Pb/HSSF	11.5	5-7 days	99.0
Ni/VSSF	20.0	72 hours	76.4
Cd/VSSF	0.0066	36 days	100.0
Cd/HSSF	0.05	5-7 days	94.0

Fe/HSSF	97.3	11 days	96.1
Mn/VSSF	0.206	79 days	100.0
Co/VSSF	20.0	72 hours	54.6
Cu/VSSF	20.0	72 hours	68.3
Cu/VSSF	0.73	30 days	100.0
Cr/VSSF	20.0	72 hours	66.2

HSSF CW: Horizontal Subsurface Flow Constructed Wetland; VSSF CW: Vertical Subsurface Flow Constructed Wetland.

Energy Requirement

Adequate water supply is important for mining operations and as water is one of key business risks for the industry, the use of sustainable technologies that treat and use water efficiently is a must. Using the LCA (Life Cycle Analysis) approach, Brix, H. compared the energy requirements of three system types namely, CWs, wastewater aquaculture facilities and conventional wastewater treatment technologies. The results revealed that the CWs require very low energy compared to the others. Austin, D. and Nivala, J. investigated the energy requirements of three wetland technologies: aerated subsurface flow, tidal flow and pulse-fed wetland treatment as engineered wetlands, and compared them to that of a mechanical activated-sludge treatment system. Their results showed that the electrical power requirements for the wetland systems were 0-56% of that of activated-sludge treatment systems. These findings agreed with a study, which used a different approach involving emergy synthesis. Emergy is described as the available energy of one kind previously required to be used up directly and indirectly to make a product or service. Shao, L., in adopting the input-output energy approach of system accounting estimated that the energy embodied in treating per m^3 of wastewater and removing per kg BOD (Biochemical Oxygen Demand), COD (Chemical Oxygen Demand), SSs (Suspended Solids) and $NH_{3-}N$ by a CW is only 6.83%, 17.48%, 13.96%, 22.47% and 9.48% respectively of that by a conventional system. Studies that compare the energy requirement of WSPs and CWs for treating mining wastewater is rare but Mara, D. D. showed that on the basis of land area requirement, capital, operating and maintenance costs, WSPs are to be preferred to SSF (Surface Flow) CW. As CWs perform better in metal removal than WSPs, it can be argued that to obtain a performance equal or better than that of a CW, more energy, land area may be required. Thus a key research question would be how much more energy would be required.

Greenhouse Gas Emissions

One of the factors to consider in assessing the economic viability of a wastewater treatment technology is its carbon footprint. Mander, U., noted that the emission factor of methane in FWS (Free Water Surface) and HSSF CWs is very high and yet the absolute value of methane emission is relatively small in the order of 1-2 magnitudes lower than that found in conventional wastewater treatment plants. They further observed that carbon dioxide emission in FWS CW was significantly lower than that of SF (Subsurface Flow) CWs while methane emission in VSSF CWs was also lower than in HSSF CWs. However, in that same study, no real differences were observed among the various types of CW in the emission of nitrousoxide. The median values for FWS, VSSF and HSSF CWs were 0.09 $mg{\cdot}m^{-2}{\cdot}h^{-1}$, 0.12 mg·m-2·$h^{-1}$ and 0.13 $mg{\cdot}m^{-2}{\cdot}h^{-1}$, respectively.

Sims, A., et al showed that a duckweed covered WSP emitted 504-1,900 mg methane and 1,700-3,300 mg carbon dioxide m-2·d-1. FWS CW resembles WSPs in its biochemical and physical processes and is therefore likely that emissions in FWS and WSPs may be similar. Studies comparing the emission rates of GHG (Green House Gas) in these treatment technologies using the same source of wastewater are rare.

GHG emissions in WSPs and CWs can be affected by various operating conditions. Methane flux properties, activities of methanogens and methanotrophs and the relationship between methane flux rate and some environmental parameters are known to be different in different plant species of CW systems. The difference in emission intensity varied with the different vegetation cover and seasonal fluctuations. In that same study, they noted that, for the same aquatic plant species, the difference in emission intensity varied with influent pollutant concentrations, the methane emission data showing large spatial and temporal variation ranging from 0 to 16.76 g CH_4 m^{-2}·day^{-1} with higher emissions occurring when influent loadings are high. Mander, U., also noted that overloading a HSSF CW led to elevated methane emissions. Further studies on GHG emissions for treating mining wastewater is necessary, taking into consideration variations created by hydrological regimes, microbial communities and plant species.

References

- The-Importance-of-Industrial-Waste-Management- 227351: johnsrefuse.com, Retrieved 14 June, 2019
- Hazardous-waste-management, technology: britannica.com, Retrieved 19 February, 2019
- Packaging-design-and-use-your-environmental-responsibilities: nibusinessinfo.co.uk, Retrieved 18 April, 2019
- Waste-management-in-food-processing-industry: bioenergyconsult.com, Retrieved 14 May, 2019
- Industrial-waste-treatment, environment-encyclopedias-almanacs-transcripts-and-maps: encyclopedia.com, Retrieved 27 July, 2019
- Composting, waste: livinggreen.ifas.ufl.edu, Retrieved 18 March, 2019
- Low-Cost-Technologies-for-Mining-Wastewater-Treatment- 322393345: researchgate.net, Retrieved 26 August, 2019

Industrial Wastewater Treatment

5 CHAPTER

Different methods of treatment of industrial waste include advanced oxidation processing, distillation, adsorption, ozonation, vitrification, incineration, chemical immobilization and landfill disposal. This chapter has been carefully written to provide an easy understanding of these treatment processes of industrial waste.

Process plants generally try to minimize the amount of wastewater they generate. However, operations invariably result in production of some wastewater. Proper treatment of this wastewater is crucial for both environmental and economic reasons.

Industrial wastewaters usually contain organic and inorganic matter in varying degrees of concentration. They may include toxic and other harmful materials as well as components that are non-biodegradable or that can reduce the efficiency of many wastewater-treatment operations.

Thus, treatment of industrial wastewaters typically is a very difficult task — far more complicated than municipal wastewater treatment —that requires special methods and sophisticated technologies. These options fall into three categories: Physical, chemical and biological. Physical treatment methods include sedimentation, flotation, filtering, stripping, ion exchange, adsorption and other processes that remove dissolved and non-dissolved substances without necessarily changing their chemical structures. Chemical methods include chemical precipitation, chemical oxidation or reduction, formation of an insoluble gas followed by stripping, and other chemical reactions that involve exchanging or sharing electrons between atoms. Biological methods rely upon living organisms using organic or, in some instances, inorganic substances for food.

Reverse Osmosis. Use of such a unit for tertiary treatment of wastewater is becoming popular.

Biological treatment is more widely used than any other option where reasonably complete treatment is required. It most often serves as the secondary treatment stage to remove major portions of contamination. Other processes handle primary and tertiary treatment to complete the removal of solids and other pollutants.

The Challenge

Some industrial wastewaters are rich in organics and easily biodegradable while others are nutrient deficient, inhibiting or preventing biodegradability. Total dissolved solids and contamination may exceed by many times the levels found in domestic sewage. Industrial wastewaters often also have pHs well beyond the range of 6–9 and may contain high concentrations of dissolved metal salts. To further complicate matters, wastewater flows and characteristics within a plant also can vary with time because of campaign manufacturing or slug discharges on top of the usual discharges. In addition, spillages and dumping that occasionally may occur very adversely can impact the performance of the plant's wastewater treatment plant. Consequently, it's always prudent to carefully assess current wastewater and its treatment requirements rather than relying on the past situation. An understanding of the nature of the plant's operations is vital.

One key parameter for wastewater is its biochemical (or biological) oxygen demand (BOD). This is the amount of dissolved oxygen needed by aerobic biological organisms to break down organic material present in a given wastewater sample at a certain temperature over a specific time period. Therefore, BOD indicates indirectly the amount of organic compounds in wastewater. The BOD most commonly is expressed in milligrams of oxygen consumed per liter of sample during 5 days of incubation at 20°C.

Another key parameter is chemical oxygen demand (COD), which indirectly specifies the amount of organic compounds in the wastewater. It indicates oxygen consumption and also is given in mg/L.

Both BOD and COD measure the amount of organic compounds in wastewater. However, COD is less specific because it measures everything that can be chemically oxidized rather than just levels of biodegradable organic matter. You can estimate the biodegradability of wastewater by considering its COD and corresponding BOD.

Primary Treatment

Removing large, suspended and floating solids is the focus of the first stage of wastewater treatment. However, before such treatment takes place, the plant wastewaters usually first go to an equalization tank or system, which acts as a buffer and normalizes varying flow and contamination loads. It's always best to use a single large concrete tank to which an appropriate coating has been applied. This tank most often is sized based on the difference between expected peak and average flows, with a capacity of 4–8 hours' worth of difference common.

From the equalization tank, the raw wastewater goes for primary treatment. This usually includes screening to trap solid objects, sedimentation by gravity to remove suspended solids and some adjustments. Primary treatment sometimes is referred to as "mechanical treatment" because it relies on mechanical methods, although chemicals often are used to accelerate the sedimentation process. The design for a primary treatment facility most often includes neutralization (i.e., pH adjustment), coagulation, flocculation and dissolved air flotation (DAF).

The main purpose of primary treatment is to remove colloidal solids, emulsified oil and a small portion of BOD and COD. Primary treatment can reduce BOD of the incoming industrial wastewater by around 20–30 % and the total suspended solids by some 50–65%.

- Neutralization: Usually, wastewater must have its pH adjusted so that subsequent operations such as downstream biological treatment can take place at optimum pH. Therefore, the wastewater passes to a neutralization system that corrects its pH. This system generally involves multiple neutralization tanks; common configurations are "3+1" (3 operating + 1 standby), "5+1" (5 operating + 1 standby) and "7+1" (7 operating + 1 standby). Injection of chemicals such as a caustic soda or sulfuric acid solution adjusts the pH to the desired level.

Sensors installed at the inlet and outlet of the neutralization tank (a minimum of one sensor in each location) measure the pH of the wastewater. A controller uses these readings to automatically adjust a dosing pump to achieve the desired final pH (typically, 6.7–8.3 with an optimum of 6.9–7.4).

The neutralization-chemical system consists of storage and mixing tanks and other equipment such as agitators necessary to reduce the concentration of the chemical and prepare it for injection. Dosing pumps are deployed in a "1+1" arrangement (1 operating +1 standby) for each chemical. Often positive displacement pumps handle these services. However, these sometimes can pose maintenance and reliability issues. A variable-speed-drive centrifugal pump often offers an attractive alternative that provides reliability and high performance.

- Coagulation and flocculation: Wastewater from the neutralization tank usually flows by gravity into coagulation tanks for removal of colloidal solids. Coagulation is a quick process, requiring a relatively low retention time of 2–5 min. There commonly are multiple rectangular coagulation tanks made of reinforced concrete with proper coating; each contains a few agitators that provide high-energy mixing. Large plants often use configurations such as "7+1" (7 operating + 1 standby) or "9+1" (9 operating + 1 standby) or similar. For instance, a treatment plant of 3,000-m^3/hr total capacity employed"7+1" tanks, each of 16 m3 capacity, to achieve retention time of more than 2.2 minutes. Some large plants have used retention times as low as 1.5 min and certain radical designs propose times as low as 1 min. However, low retention times can pose risks. Generally, it's wise to keep retention times above 2 min.

A coagulant solution (typically polymer based) usually is injected automatically by dosing pumps ("1+1" configuration); most often stroke variation adjusts injection. Modern plants automatically control injection rate according to incoming flow rate based on a more or less fixed chemical concentration, preliminarily defined through site experimental tests and adjustable during normal plant operation. Coagulant aid can be added to the wastewater stream to facilitate separation of solids.

Wastewater from coagulation tanks most often flows by gravity into the flocculation system (tanks) where agglomeration of flocculent formed during coagulation process takes place. Anionic polymer usually serves as flocculent. Flocculation is a process of slow mixing with retention times of 12–40 min. Some designs for large plants have used lower retention times, say, 9–10 minutes, but typically times of 11, 12 or 15 min. are recommended. It is a process that requires less energy for agitation than coagulation.

- Dissolved air flotation: Wastewater from flocculation passes by gravity into a DAF clarifier system. Its main purpose is to remove the suspended solids, emulsified oil, grease and some portions of BOD and COD from the wastewater. Elimination occurs through the action of micron-sized air bubbles. These are created by dissolving air in wastewater under pressure and then reverting to atmospheric pressure in DAF clarifiers. The millions of micron-size air bubbles released attach to the contaminants, decreasing their effective density and thus causing them to float on the surface to form a concentrated sludge blanket. A skimming device removes the floating sludge, which then go to sludge treatment units for processing. A common design uses a separate pressure vessel for compressed air introduction. DAF clarifiers operate effectively over a wide range of hydraulic and contamination loading.

Secondary Treatment

Often considered the heart of the treatment plant, its major purpose is to remove biodegradable organics (expressed as BOD, COD, etc.) and ammonia. Secondary (or biological) treatment uses microbes to consume dissolved organic matter that escapes primary treatment, converting it to carbon dioxide, water and energy for microbe growth and reproduction. After this biological process, the stream goes to additional settling tanks ("secondary" clarifiers or sedimentation vessels) to eliminate more of the suspended solids. Well designed and functioning secondary treatment can remove about 85–90% of the suspended solids and BOD. Technologies employed include the activated sludge process, which is the most commonly used method, as well as variants of pond and constructed wetland systems, trickling filters and other forms of treatment that rely on biological activity to break down organic matter.

An activated-sludge train usually is divided into an aeration section for BOD removal and nitrification, and an anoxic section for denitrification. In the aeration section, compressed air passes through the wastewater. Dissolved oxygen from the compressed air acts as a respiratory source for aerobic bacteria present in wastewater that decompose the organic load (expressed as BOD and COD) and ammonia to carbon dioxide and nitrates, respectively.In the anoxic section, bacteria use the oxygen in nitrates as a respiratory source, thus converting the nitrates to nitrogen gas.

In practice, denitrified wastewater from the anoxic tank flows downstream to the aeration (or BOD-removal) tank where aerobic bacteria decompose the organic load and ammonia present using dissolved oxygen supplied by air blower(s). The treated effluent from the aeration tank usually flows by gravity to a secondary clarifier, which most often is a gravity clarifier. Here, sludge is removed from the treated effluent, which then passes to tertiary treatment. A portion of the sludge gets recycled to the anoxic section to provide nitrates for denitrification. This recirculation keeps effluent nitrates' concentration below the required limits. The remaining portion of sludge goes to sludge treatment facilities.

Biological treatment usually consists of multiple streams, say, 4, 6 or 8 trains, with a proper safety factor (for instance, 1.5 or more) to ensure the biological treatment can handle the incoming design flow even if one train is taken out of operation. Selection of the hydraulic retention time for the anoxic zone requires great care. Considering different operational and process factors, as a rough indication, this time usually is 5–8 hr. Some designs for large plants have used 5.5 hr, 6 hr and 6.5 hr as optimum values. Hydraulic retention time for the aeration tank is longer, somewhere between 19 and 24 hr. Some large plants have found a retention time of 20 hr to be optimum.

Tertiary Treatment

This ensures removal of remaining contamination and solids in the wastewater. Such tertiary treatment usually involves filtration systems such as disc filters, reverse osmosis (RO) units, etc. You usually should direct filter reject from backwash or RO unit rejects to the flow distribution chamber upstream of the equalization system; most often, these rejects require a dedicated pumping system. To eliminate specific contaminations to meet regulatory requirements, many plants must resort to special treatment, e.g., the Fenton process to remove non-biodegradable COD. While other technology options are available, the Fenton process most often is selected because of its reliability, initial cost, operational cost and footprint.

The Fenton section usually consists of dosing systems for hydrogen peroxide and ferrous sulfate. After dosing with chemical in an oxidation tank, the wastewater goes to tube settlers to settle out the contaminants. During regular operation plants generally don't need to put wastewater through such treatment. However, having a Fenton section can ensure treatment adequacy when facing sustained peak COD in the wastewater.

Many units in tertiary treatment such as that for the Fenton process or fine filtration should consist of multiple parallel streams to provide flexibility during operation. Commonly used arrangements are "n+1" and "n+2" — for example, "2+2" "3+1" "4+2" and "5+1".

Effect of Wastewater

Both bodies of freshwater and saltwater are polluted every day by untreated wastewater. In fact, the U.S. EPA estimates that almost 1.2 trillion gallons of sewage from household and industrial sources is dumped into the nation's water every single year, or about 3.28 billion gallons a day. And, yes, that was trillion with a "T" and billion with a "B". This not only creates an unsafe environment for marine life, it creates hazards for humans as well. The importance of wastewater treatment design and infrastructure is especially relevant with bodies of fresh water, as these materials would end up in your home if water wasn't treated properly.

Groundwater and Water Tables

Numerous parts of the world are currently suffering from water scarcity (and that includes U.S. states like California), which means clean water is of the utmost importance. When wastewater is discharged on these dry lands, it can seep into the underground water tables and well sources. Because we need to draw from these natural bodies of water for generations to come, this can render entire water supplies useless for people in multiple locations.

Natural Ecosystems

Every ecosystem relies on water in some regard. And when water is contaminated by sewage, toxic chemicals, or any number of other man-made forms of waste, those ecosystems are put at serious risk. Not only that, but surface and underground water are connected, always. Reckless disposal of waste can contaminate a far wider range of animals and environments than you may even know.

Organica Way

Current wastewater treatment technologies may help prevent some of these issues, but they also come at the cost of large, industrial facilities that can have adverse environmental effects themselves. At Organica Water, we believe there's a better way. Our facilities utilize live biofilms on plants or engineered structures inside a fully enclosed area. Unlike most water treatment plants there are no bad smells or eyesores, which means they can be built even in densely packed urban areas. Most importantly, this unique wastewater treatment design results in a cleaner and greener wastewater management plant.

TREATMENT OF INDUSTRIAL EFFLUENTS

Primary Treatment of Industrial Effluents

It is of general nature and is used for removing suspended solids, odour, colour and to neutralize the high or low pH.

It involves methods of:

- Screening.
- Neutralization.
- Equalization.
- Sedimentation.
- Coagulation.

Some general treatment processes are given below:

Screening

It is a process through which large materials like wooden pieces, metal pieces, paper, rags, pebbles, fibres etc. are removed. The rotary and circulation filters are used now a days in modern industries to remove large materials.

These both methods are effective and help in reducing suspended solids and BOD of the industrial effluent. The micro strainer is also used to remove five suspended particles in some treatment processes. To remove, colloidal matter, ultra filters are also used although they are costly.

Neutralization

When pH of the industrial waste is too high or too low then it should be neutralized by acid or alkali and only neutral effluent should be discharged into the nullah or public sewer.

Lime Stone Treatment

For acidic effluent, lime stone should be used as it will form calcium compounds [$CaCl_2$, $CaBr_2$, $Ca(NO_3)$ or $CaSO_4$] depending upon the presence and amount of acid.

Caustic Soda Treatment

Although it is costly method but it is also utilized for neutralizing the acid. Here caustic soda is added in the effluent to make the pH neutral. Only small amount of caustic soda is needed for this work.

For neutralization of alkaline effluent the following techniques are used.

Carbon Dioxide Treatment

If factory is producing carbon dioxide then only this method should be utilized for neutralizing the pH otherwise it would be costlier affair. Here CO_2 is passed in alkaline effluent to make its pH almost 7.

Sulphuric Acid Treatment

This is a common method of neutralizing alkaline effluent. Here sulphuric acid is added in the effluent till pH becomes almost 7.

Utilizing Waste Boiler: Flue Gas

The stack gas which contains about 12% carbon dioxide is utilized to react alkaline effluent to make it neutral.

Equalization

When effluent is discharged from factory then its pH along with the quantity of suspended solids, dissolved solids etc. vary from the beginning to the last depending upon the dilution, velocity and the amount of reactants etc.

Hence as the character of the effluent do not remain the same throughout hence proper treatment is not possible. So equalization tank is necessary where effluent is keep for 10 hrs or more for the stabilization of pH and BOD. During equalization suspended solids settle down and new acid of alkaline treatment becomes economical.

The equalization tank should of sufficient size so that it may retain even the effluent of the whole day. Generally rectangular basins are selected for this purpose. If any how the arrangement for mechanical agitation is also done for some time in the tank then separation of suspended particles becomes more easier.

Sedimentation

This treatment is only employed for the settlement of suspended particles by gravity. This technique is only used in the beginning to settle down the solid particles in a high suspension effluent.

Coagulation

Experimental results have shown that a slit particle of size 0.05 mm requires about 11 hours to settle down through a depth of 3 m and clay particles of size 0.002 mm require about 4 days' time to settle the same height of 3 m of at normal temperature of about 25°C. As we know that water contains colloidal impurities which are even finer than 0.0001 mm and which also carry electrical charge on them.

Due to electrical charges they remain in motion and never settle down. Therefore when water is turbid due to presence of such fine size and colloidal impurities, plain sedimentation is of no use. It is also not possible to provide detention periods of longer than 4 — 9 hours. The coagulation becomes necessary when the turbidity is more than 40 — 55 ppm.

For dealing waters with such impurities a chemical process was evolved. This process removes all these impurities within reasonable period of 3 — 4 hours. This chemical process is called coagulation and the chemical used in the process is called coagulant.

Principle of Coagulation

The principle of coagulation can be explained from the following two aspects:

- Floe formation.
- Electrical charges.

Floe Formation

When coagulant is added to the water and thoroughly mixed, it produces a thick insoluble gelatinous precipitate. This precipitate is called floe. The floe has the property of arresting the suspended impurities in water during its downward settlement towards the bottom of the tank.

The gelatinous precipitate has therefore the property of removing fine and colloidal particles quickly. The coagulation process also removes colour and test in general.

Electrical Charges

The flock ions are electrically charged (positive) while all the colloidal particles have negative charge. Therefore floes attract the colloidal particles and cause their removal easily by settlement at bottom of the vessel in which it is used.

Coagulants

The chemicals given below can be used as coagulants either alone or in combination:

- Sodium aluminate.
- Sodium aluminate + Aluminium sulphate.
- Aluminium sulphate.
- Sodium aluminate + Ferric chloride.
- Aluminium chloride (but used under exceptional circumstances only).
- Aluminium sulphate + Caustic soda.
- Ferric chloride alone.

- Aluminium sulphate + Hydrated lime.
- Polyelectrolytes.
- Ferrous sulphate.
- Copper sulphate.
- Sodium aluminate + Magnesium chloride.
- Copper sulphate + Hydrated lime.
- Ferric sulphate.
- Aluminium sulphate + Sodium carbonate.
- Ferric sulphate + Hydrated lime.
- Ferrous sulphate + Hydraed lime.
- Ferrous sulphate + Chlorine.
- Potassium permanganate + Ferrous sulphate.
- Magnesium carbonate.

In water treatment plants following are the usual coagulants most commonly used:

- Ferrous sulphate and lime.
- Magnesium carbonate.
- Polyelectrolytes.
- Aluminium sulphate.
- Sodium aluminate.
- Chlorinated copperas.

Characteristics of these six coagulants are given below:

Ferrous Sulphate and Lime

When lime and ferrous sulphate are added to water, the ferrous hydroxide is formed in the form of flocs.

$$FeSO_4{\cdot}7H_2O + Ca(OH)_2 \rightarrow Fe(OH)_2 + CaSO_4 + 7H_2O$$

$$4Fe(OH)_2 + O_2 + 2H_2O \rightarrow 4Fe(OH)_3$$

Ferric hydroxide Fe $(OH)_3$is a floe which settles down and removes all the colloidal impurities in water. Ferrous sulphate is most effective in pH range of 8.5 and above.

Magnessium Carbonate

It removes turbidity and colour.

Floes formed are heavier than formed by other processes and thus detention period is very much reduced.

It removes iron and manganese completely.

It is possible to recycle and reuse the coagulant by passing the sludge through water containing carbon dioxide as follows:

$$Mg\,(OH)_2 + CO_2 \rightarrow MgCO_3 + H_2O$$

Polyelectrolytes

Polyelectrolytes are special types of polymers. They may be anionic, cationic and non-ionic, depending upon the charge they carry. Out of these only cationic polyelectrolytes can be used independently, while other types can be used only along with other conventional coagulants.

Separam, wisprofloc, Mogul etc. are some of its patented forms available in foreign countries extract from seeds of a plant named Nirmali can act as polyelectrolyte. The plant is found in Andhra Pradesh, Madhya Pradesh, West Bengal and Orissa.

Aluminium Sulphate

It is also known as alum or filter alum. Its chemical composition is $Al_2(SO_4)_3.18H_2O$. It reacts with alkaline water to form aluminium hydroxide (floe), calcium sulphate and carbon dioxide.

The following are the chemical reactions which alum performs with various types of alkaline substances:

$$Al_2(SO_4)_3.18H_2O + 3Na_2CO_3 \rightarrow 2Al(OH)_3 + 3Na_2SO_4 + 18H_2O + 3CO_2$$
$$Al_2(SO_4)_3.18H_2O + 3Ca(HCO_3)_2 \rightarrow 2Al(OH)_3 + 2CaSO_4 + 18H_2O + 6CO_2$$
$$Al_2(SO_4)_3.18H_2O + 3Ca(OH)_2 \rightarrow 2Al(OH)_3 + 3CaSO_4 + 18H_2O$$

Alum is in most common use due to following reasons:

- In addition to turbidity it also reduces taste and odour.
- It produces clear water. In other words, it is very efficient type of coagulant.
- Floes formed by it are more stable and heavy than that formed by other coagulants.
- It is not harmful to health.

Sodium Aluminate

Chemical equation may be as given below:

$$Na_2Al_2O_4 + CaSO_4 \rightarrow CaAl_2O_4 + Na_2SO_4$$
$$Na_2Al_2O_4 + CaCl_2 \rightarrow CaAl_2O_4 + 2NaCl$$
$$Na_2\,Al_2O_4 + Ca(HCO_3)_2 \rightarrow CaAl_2O_4 + Na_2CO_3 + CO_2 + H_2O$$

The effective range of pH for this coagulant is 6.0 to 8.5.

Chlorinated Copperas

The combination of Ferric sulphate $Fe_2(SO_4)_3$ and Ferric chloride $FeCl_3$ is called chlorinated copperas. When solution of ferrous sulphate is mixed with chlorine, both Ferric sulphate and Ferric chlorides are produced. About 1 kg chlorine reacts with 7.3 kg of ferrous sulphate.

$$6FeSO_4.7H_2O + 3Cl_2 \rightarrow 2Fe_2(SO_4)_3 + 2FeCl_3 + 42H_2O$$

Ferric sulphate and Ferric chloride each is an effective floe and their combination is also very effective.

It is found that Ferric chloride and Ferric sulphate can be used independently with lime to act as coagulants. In this case following are the chemical reactions.

$$Fe_2(SO_4)_3 + 3Ca(OH)_2 \rightarrow 3CaSO_4 + 2Fe(OH)_3$$

$$2FeCl_3 + 3Ca(OH)_2 \rightarrow 3CaCl_2 + 2Fe(OH)_3$$

Ferric chloride is effective in pH range of 3.50 to 6.50 or above 8.50 but ferric sulphate is effective in pH range of 4 to 7 and above 9.

Dry Fed Devices

There are a number of devices which may be used for dry feeding. But two most commonly used devices are shown in figure. Dry powder of coagulant is filled in the conical hopper. The hoppers are fitted with agitating plates which prevent the chemical from being stabilized. At the bottom of the hopper a revolving helical screw or the toothed wheel is fixed.

The rotation of the helical screw or the toothed wheel is regulated through a venturi device in the raw water pipe. When more discharge is passed through the venturi device, the rotation of the screw or toothed wheel gets increased and more coagulant is thrown in the water.

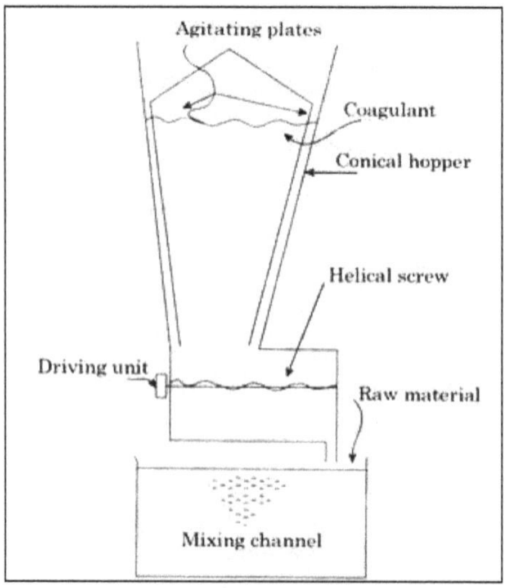

Dry fed device.

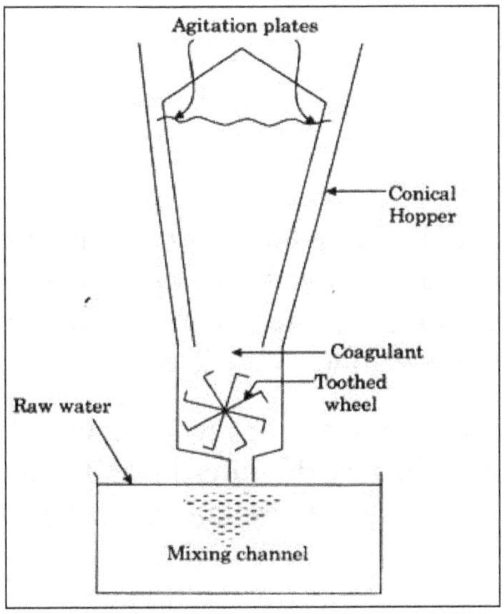

Anode dry fed device.

Mixing Channels

In this method, the mixture of raw water and coagulant is made to pass the channel in which fluming is done. After fluming vertical baffles are also fixed on both the sides of the channel. The complete arrangement has been shown in figure.

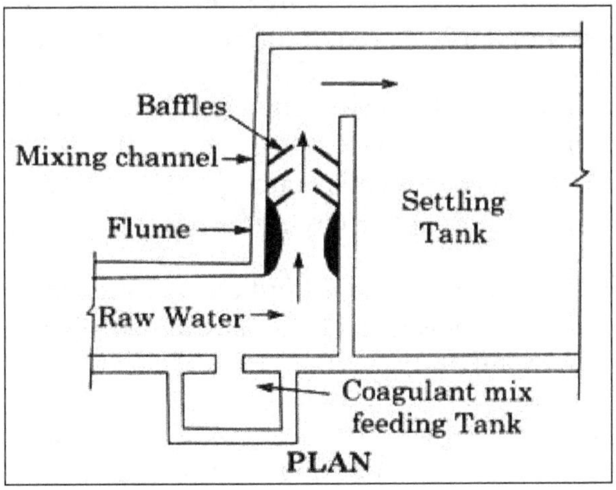

Mixing channel. Separation column.

Secondary Treatment of Industrial Effluents

The main biological treatments are given below:

- Trickling filter.
- Anaerobic digestion.
- Oxidation ditch.
- Aerated lagoon.

- Activated sludge process.
- Oxidation pond.

Trickling Filter

A trickling filter, also known as percolating filter or sprinkling filter, is an artificial bed of stone or broken brick material over which waste water or sewage is allowed to sprinkle or to trickle. It is then collected through the under drainage system. A zoolial film is formed on the filter media and oxidation of organic matter takes place under aerobic conditions.

A bacterial film, known as a bio film is formed around the particles of filtering media and for the existence of this film, oxygen is supplied by the intermittent working of the filter and by suitable ventilation facilities in the body of the filter. The effluent is collected in the under-drainage system and ventilated.

The filter bed is provided with ventilators along the entire periphery of the film at 2m centre to centre. They are raised above the media top by 75 mm and covered with cowl. The filter media consists of crushed stone, gravel, slag, broken brick blocks of inert materials etc. These are placed either in a single or multilayers.

The colour of this film is blackish, greenish and yellowish and it consists of bacteria, fungi, algae, lichens, protozoa etc. The trickling filters are broadly divided as standard rate trickling filters and high rate trickling filters. The standard rate trickling filters have a hydraulic loading (total volume of liquid applied per hour per unit surface area of filtered bed) of 525 to 2100 m/h/per hectare.

The high rate trickling filters have a hydraulic loading of 4200 to 15000 m^3/h/ hectare. The organic loading of standard rate trickling filters varies from 80 to 400 g/day/m^3 and that of high rate trickling filters varies from 400 — 4800 g/day/m. Organic loading is the total weight of 5 — days at 20°C BOD applied per day per unit volume of filter media. Organic loading is generally expressed as g/day/m^3.

In the standard rate trickling filter the sewage is applied intermittently by means of dosing tanks. In the high rate trickling filter the application of waste water or sewage is continuous and treated or partially treated sewage is recirculated.

Thus a portion of the treated waste water is returned to the treatment process. The return may be made after the secondary tank or to the raw sewage of the primary settling tank or to the dosing tank of the filter.

High rate trickling filters thus operate in two or three stages with or without sedimentation between these stages. In other words, the effluent is circulated twice or thrice with addition filters in accordance with quality of treated effluent needed.

The efficiency of high rate trickling filters is greater than the standard rate trickling filters. Trickling filters may be circular or rectangular in shape, but circular form is more common.

The rate of filter loading (kg of BOD per volume of filter bed) varies from 1000 to 2200 kg of BOD per hectare — metre per day. The rate of filter loading (kg of BOD per volume of filter media) varies from 15 to 30 kg of BOD per day per 100 m^3 of filter material.

The rate of filter loading (surface area of filter bed) varies from 25 to 40 million litres per hectare of surface area per day. The rate of filter loading (volume of filter bed) varies from 7.50 to 22.50 million litres per hectare — metre per day.

Now a days instead of stones, plastic and PVC circular pieces are used being lighter and economical. The depth of bed is generally taken about 2.5 to 3 metre depending upon the size of stone, temperature of effluent and retention time. As the function of organisms is directly proportional to temperature decrease the efficiency of the filter.

According to bio-chemist the bacteria includes species of genera, pseudomonas, Alcaligens, Micrococcus, Flavo bacterium and Entero bactericeae. It has been observed that during the treatment Fungi and algae are also present in trickling filters. Thus it is a mixture of various organisms which is responsible for the oxidation of organic matter present in the effluent.

Anaerobic Digestion

This treatment is in fact slow oxidative digestion process carried out in absence of air a closed container where ammonia and methane are released as the end products of the reaction. A very important point to be taken under consideration is that pH must be maintained almost 7 during the oxidation process.

In this treatment bacteria desulfo vibrio acts in reducing sulphates to sulphides and H_2S is liberated. Besides this, bacteria belong to genera, methanosarcina, methano bacterium and methanococcus act to produce methane. The bacteria flavobactrium alcaligenes, pseudomonas, aerobacter and escherichia act to produce acids.

This treatment is generally used in paper mill, dairy industry, slaughter house and other factories which produce soluble organics.

Oxidation Ditch

It is a treatment generally carried out in beet sugar manufacturing plants, diary, slaughter house and other factories to reduce BOD to 85 to 95%.

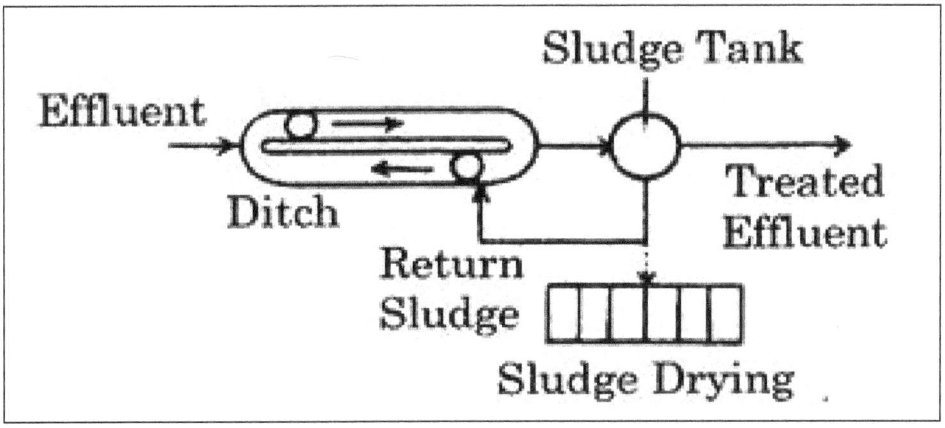

The oxidation ditch is an improvement over activated sludge process in shape and the aerator. The aerator is cage rotor placed across the channel.

Oxidation Ditch Process

Oxidation ditch is a long continuous channel (lined with butyl rubber or plastic) generally oval in shape and about 1 to 2 metre in depth. The effluent after primary treatment is passed into oxidation ditch and kept here for long time as it is a slow process. High efficiency can be obtained by recycling about 98% sludge in such a fashion that overall sludge production is lowered down.

Aerated Lagoons

These are big cement tanks having a depth of 4-6 metres. These tanks or lagoons are used for the oxidation of dissolved organics.

When the waste of the factory is passed in these lagoons and aerated mechanically then after 3-4 days (depending upon the quantity and temperature) flocculent sludge is formed which is responsible for the oxidation of soluble organic material present in the effluent? If proper precautions are taken then 85-95% BOD is reduced with aerated lagoons.

The only drawback of this method is that it requires a long space for tanks.

Activated Sludge Process

It is an important biological oxidation method for the removal of suspended and colloidal solids and also reduces BOD of the effluent. In this method, effluent is continuously subjected to biological degradation carried out by microbial suspended in the reaction tank into which oxygen is introduced by mechanical methods.

The effluent which comes after reaction tank is allowed to settle down and a portion of sludge is recycled to the tank itself for microbial population.

The process diagram is given below:

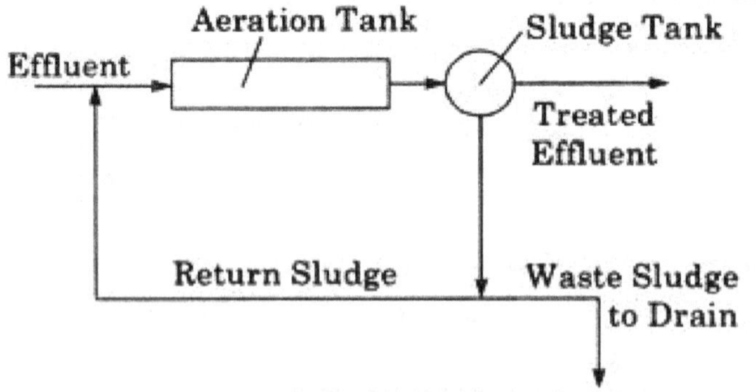

Activated Sludge Process

The important factors which determine the efficiency of the activated sludge are as follows:

- PH value.
- Temperature.
- Volume of tank.

- Nature of organic matter.
- Velocity.
- Oxidation: Reduction potential of sludge.

The microbial mass consists of bacteria, fungi, protozoa, nematodes and rotifers, which are formed in activated sludge process and are the active agents of activated sludge. The micro-organisms are supplied by adding essential nutrients mainly nitrogen and phosphorus. Generally Nitrogen and Phosphorus are supplied by addition of urea and mono ammonium or diammonium hydrogen phosphate.

The nitrogen and phosphorus requirement are about 10% and 2% per day. The actual amount can be calculated on the basis of the quantity of the sludge. The researches have shown that if trace amount of iron, cobalt and molybdenum are added then the efficiency of the oxidation increases.

The optimum pH for activated sludge should be between 6.5 – 9.0 as below 6.5 fungi will complete with bacteria and above 9.0 rate of metabolism will be reduced. For maintaining this range of pH a buffer can also be added (Bicarbonate can be added to maintain pH).

In this technique as pH is a very important factor hence continuous monitoring of pH, is necessary during the treatment. Besides this high temperature is also needed as it enhances metabolic activity due to which oxygen is consumed in a quicker time and thus anaerobic conditions are attained immediately.

Some chemists have suggested the following modifications in this process:

- Contact stabilization.
- Tapered aeration.
- Extended aeration.
- Tapered aeration.

Out of these aeration techniques, extended aeration is preferred where only aeration period is extended (from 10—12 hrs) to 30 — 45 hrs. If the effluent is passed at several places of the aeration tank then it is called 'Stepped Aeration' while reduction of oxygen supply continuously in the tank is called 'Tapered aeration'.

Now a days a modified activated sludge process called 'High-rate aerobic treatment' is used. In this technique, aeration period is extended from 30 to 70 hrs, settling and returning of settled sludge to the aeration tank. Here a big tank is taken so that complete aeration period may be given to the effluent.

Oxidation Pond

It is a pond where oxidation takes place with the help of bacteria- pseudomonas, alcaligenes and flavo bacterium and algae. In fact the presence of bacteria and algae both help in the oxidation of organic load of effluent.

The dimensions of the pond depend upon the quantity of the effluent release per day. The depth of the pond should be between 1.5 metre — 2 metre otherwise it would be come breeding place for mosquitoes and midges. Preference has been given for the depth of 2 metre as in this case pond becomes anaerobic rather than aerobic.

The bottom of the pond and all areas should be made of concrete or brick to avoid seepage of sludge etc. through the pores. Polythene sheeting should be spread in the lower portion to avoid any chance of seepage through any side. The tank should be made in such a way that at the entrance of the placement it should be more deeper that at the outlet to create anaerobic zone for solid deposition and digestion.

The mechanism of the pond reaction is as follows: when the effluent is added in the pond the solids slowly and slowly settle in the bottom and this solid layer acts as anaerobic phase and anerobos act to convert organic matter into ammonia, methane and carbon dioxide. In the top layer of the pond the oxidation takes place releasing the products, water and carbon dioxide.

The facultative zone is near anaerobic phase hence the presence of excess alage in the pond will enhances the rate of oxidation of the organic matter. In fact the oxygen need for the metabolism is given by the algae in the pond and which in turn absorbs carbon dioxide released by bacteria from photosynthesis.

Thus both bacteria and algae help in the oxidation of the organic matter of the pond and hence their presence in good amount is necessary for the effective purification of the effluent. As we know that for photosynthesis of algae the sun light is necessary hence precautions should be taken in such a way that maximum light penetration should occur in the pond.

The following factors are important for the effective anaerobic and aerobic oxidation of the pond:

- The pond should be open and larger and depth of pond should be 2 metre.
- Sun Light penetration in the pond for photosynthesis of algae.
- Wind action for mixing.
- Nutrient level should be neither too high and nor too low as in case of high euglena and chlorella dominate and for low nutrient level filamentous algae such as ulothrix, vaucheria and spirogyra will be generated.
- The pH of the effluent should be high as it will lead to precipitation of the heavy metals as hydroxides which settle as sludge.
- For disinfection of waste material, it should be taken into another pond or disinfectant should be added separately. Disinfectant should not be added in the oxidation pond otherwise the whole mechanism will be stopped.

Tertiary Treatment of Industrial Effluents

This type of treatment is needed for the effluent for the removal of bacteria and dissolved inorganic matter (metals, metal oxides, metal carbonates, metal sulphates etc.

The bacteria of fecal origin in remove by keeping the effluent in maturation ponds for a definite period. If any how there are still bacteria in the effluent then bleaching powder is added to kill them or a higher dose of chlorine gas is passed for a definite interval of time say 1 to 1.5 hours.

According to latest researches conducted by V.P. Kudesia and Co-Workers if a mixture of chlorine and bromine solution is passed in the effluent then it kills all type of bacteria within 30 minutes while if chlorine of same concentration takes 1 hour.

This reduction of time is due to the fact that a mixture of 4 species of chlorine (CI_2, HCl, CI_3^- and Cl^+) and 4 species of bromine (Br_2, HOBr, Br_3^- and Br^+) when combine lead to formation of several new species along with more formation of molecular chlorine and molecular bromine and which cause the rupture of the bacterial cell.

For the removal of inorganic substances, the following methods have been recommended:

- Reverse osmosis.
- Chemical Precipitation.
- Evaporation.
- Dialysis.
- Removal by algae.
- Activated carbon.
- Exchange resins.

Reverse Osmosis

The passage of solvent from a solution of low concentration to high concentration through semi permeable membrane is called osmosis. The reverse of this process is called reverse osmosis.

Here effluent containing dissolved solid is passed through semipermeable membrane at the pressure excess of osmotic pressure of feed waste (about 45-50 atm) the water of the effluent passes through the semi permeable membrane leaving the dissolved solids on the surface. Thus through this technique the solid inorganic matter is separated from the water and the material can be recovered by using proper techniques.

Although this method is costly as collulose acetate or polyamide hydrazide are used as semipermeable membranes but it is an effective method and yields immediate results.

Chemical Precipitation

It is a chemical technique by which metals are removed by precipitating them either as hydroxides at high pH or as sulphates etc. For example, chromate from electroplating industry can easily be removed by reaction with ferrous sulphate and then precipitating it with lime.

From Barium chloride factory, the excess of Barium is precipitated as Barium sulphate by the addition of Dil. Sulphuric acid. The nickel can be precipitated with the help of dimethyl glyoxime and mercury with the help of potassium iodide as mercuric iodide. Thus by use of suitable reagents, metals can be precipitated and separated and reutilized.

Evaporation

This method is generally employed when waste solid/solids are reused in the industry. It is a method used for recovery of radioactive substances. Here the effluent is boiled and after the evaporation of the water, the concentrated solution is left out in the vessel which is again used in the recycle process of the industry.

Dialysis

The process of separating crystalloid from colloid by diffusion or filtration through a membrane is called dialysis.

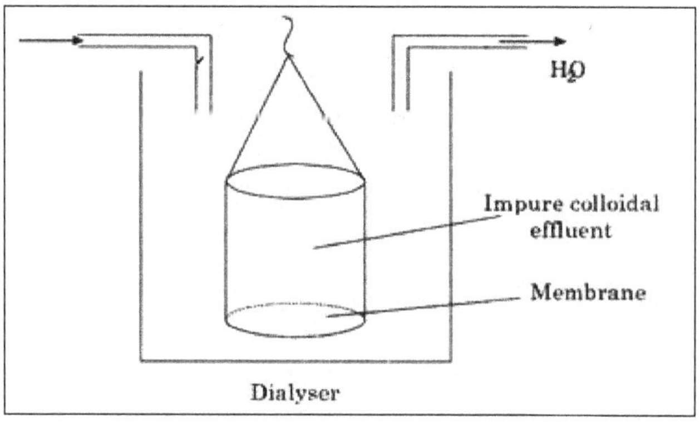

The dialyser consists of a shallow cylinder open at both ends, over one end of which is membrane is tied. The effluent which is to be dialysed is placed in this cylinder which is then suspended in a large dish containing water. The flow of water is continuous. Dialysis is a slow process and takes several hours for the complete removal of the substances.

Take for example the effluent containing hydrochloric acid, and sodium silicate then resulting sodium chloride and hydrochloric acid diffuse away, leasing only colloidal silicic acid in the membrane.

Removal by Algae

As we know that algae require metals — cobalt, copper, zinc, manganese iron, molybdenum etc. in trace amounts and potassium, calcium magnesium, phosphorus, nitrogen and sulphur for the growth hence presence of algae in the effluent will reduce the above contents from the effluent. It is important to note that algae must be removed prior to its reuse otherwise it would create problems.

Activated Carbon

As activated carbon has high adsorption power hence it is utilized for the removal of pesticides such as DDT, hexachlorobenzene, Dieldrin, Heptachlor, Lindane, Aldrin, Chlordane, Toxophene, Methoxychlor, Heptachlor epioxide and others. The simple mechanism involves in this techniques is based on the phenomenon of adsorption.

As adsorption is a surface phenomenon hence the pesticides or other substances get adsorb on the surface of the activated carbon and can easily be removed and reutilized. The advantage of this technique is that adsorbed substances can be recovered in the same form in which they were present.

Resins

Generally, ion exchange resins are insoluble in water and in organic solvents and they contain active or counter ions that will exchange reversibly with other ions, in a surrounding solution without any appreciable physical change occurring in the material.

The ion exchanger is of complex nature and is polymeric. In a cation exchanger the active ions are cations while in an anion exchanger, active ions are anions. The structure cross-liked sulphonated polystyrene is shown below.

It is said that when resin is brought in contact with water, some water penetrates into the resin, and the hydrogen atoms of the sulphonic acid ionize. They may then be replaced by an equivalent quantity of another cation.

Cation Exchange Resins

They have a high molecular weight, having cross-linked polymer containing sulphonic, carboxylic, phenolic etc. groups as an integral part of the resin and an equivalent amount of cations.

Anion Exchange Resins

They are also polymers containing amine or, quaternary ammonium groups as integral parts of the polymer lattice and an equivalent amount of anions such as chloride, hydroxyl or sulphate ions.

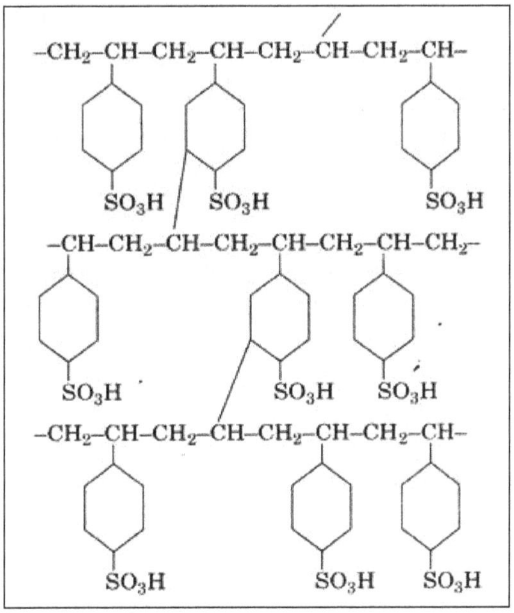

Characteristics of a useful resin are:

- The resin must be sufficient cross-linked to have only a negligible solubility.

- The resin must contain a sufficient number of accessible ionic exchange groups and must be chemically stable.
- The resin must be sufficiently hydrophobic to permit diffusion of ions through the structure at a finite and unstable rate.
- The Swollen resin must be denser than water.

Action of Ion-exchange Resins

It is clear that cation exchange resins contain free exchangeable cations.

$$\left(\text{Res.}\,A^-\right)B^+ + C^+(\text{so In.}) \leftrightarrow \left(\text{Res.}\,A^-\right)C^+ + B^+(\text{so In.})$$

where (Res. A^-)B^+ represents the cation-exchange resin in which resin is the basic polymer of the resin. A^- is the anion attached to the polymeric frame work and B^+ is the active or mobile cation.

Hence a sulphonated polystrene resin in the hydrogen form as (Res. SO_3^-) H^+. Similarly an anion exchange resin may be written as (Res. NMe_3+)Cl-.

If we observe that the equilibrium is completely displaced from left to right, the ions C_3^+ is completely fixed on the cation exchanger.

If the solution contains several ions (C^+, D^+ and E^+) the exchanger may show different affinities for them thus making separation possible. We here consider the displacement of sodium ions in a sulphonate resin by calcium ions.

$$2\left(\text{Res.}\,SO_3^-\right)Na^+ + Ca^{++}(\text{Soln.}) \leftrightarrow \left(\text{Res.}\,SO_3^-\right)_2 Ca^{++} + 2Na^+(\text{soln.})$$

Since the reaction is reversible, hence the resin may be recovered by passing a solution containing sodium ions through the product, the calcium ions are thus removed from the resin.

Cation Exchange Resins

Consist of small, simple cation and can be exchanged by large, high molecular weight anions.

$$\text{Resin} - H + NaCl \rightarrow \text{Resin} - NaHCl$$

Anion Exchange Resins

Consist of small, simple anions which can be exchanged by large, high molecular weight cations.

$$\text{Resin} - OH + NaCl \rightarrow \text{Resin} - CI + NaOH$$

Both natural and synthetic materials are available as ion exchange resins.

The factors determining the distribution of ions between an ion-exchange resin and a solution include.

Nature of Exchanging Ions

- The extent of exchange increase with increasing valency of the exchanging ions (At low concentration and at ordinary temperatures), e.g.,

  $Na^{+} < Ca^{++} < Al^{+++} < Th^{++++}$

- The extent of exchange increase with decrease in size of the hydrated cation (under similar conditions and constant valence).

  $Li^{+} < H^{+} < Na^{+} < NH_4^{+} < K^{+} < Rb^{+} < Cs^{+}$

- When a cation in solution is being exchanged by an ion of different valency the relative affinity of the higher valent ion increase in direct proportion to the dilution.
- With strongly basic anion exchange resins univalent anions appear to behave similarly to univalent cations.

Nature of Ion-exchange Resin

The absorption of ion will depend upon the nature of the functional groups in the resin.

The result obtained in ion-exchange separations can be affected by:

- The solvent or eluent.
- Temperature.
- Particle size.
- Nature of the ion-exchange resins.
- The hydrogen ion concentration.
- The length of the column.
- The rate of flow of eluant.

The important properties which notice the behaviour of a resin are:

- Number of functional groups.
- Nature of functional groups.
- Degree of cross linking.
- Size of particles.
- Strength of functional groups.

Theoretical Principles

Ion Exchange Equilibria

The net result of an ion exchange reaction may be expressed as a replacement of equivalent quantities of like charged ions.

$$RCl + OH^- \rightarrow ROH + Cl^-$$

$$HR + Na^+ \rightarrow NaR + H^+$$

$$2HR + Ca^{+2} \rightarrow CaR_2 + 2H^+$$

$$2HR + Mg^{+2} \rightarrow MgR_2 + 2H^+$$

Where, R^+ or R denotes the resin matrix. In applying the law of mass action on,

$$RCl + OH^- \rightarrow ROH + Cl^-\ ,$$

we have:

$$K = \frac{a_{ROH} \cdot a_{Cl}}{a_{RCl} \cdot a_{OH}}$$

The value of the equilibrium constant or selectivity coefficient K remains constant if activities of various species are used — otherwise K will change due to change in activity coefficients.

BIO TREATMENT OF INDUSTRIAL EFFLUENTS

Biodegradation of Pollutants

Biological treatment of effluents is a long-established practice in many countries, but some constituents of these effluents are calcitrants and, thus, not amenable to conventional treatments. Biotechnology helps in overcoming this problem.

Following example will illustrate the use of biotechnology in biodegradation of pollutants:

- In USA a company is using lignin degradation for treatment of substances like polychlorinated biphenyls and ditoxin.
- In Europe, companies are working on enzymatic detoxification to breakdown substances viz., cyanides, and also the by-products from the synthesis of S-triazine herbicides.
- Microbial transformation of biarylethers, cyclic biarylethers, biarylketones, halogenated bibenzodioxins and dibenzofurans are used to avoid the problem of release of effluents with pollutants. For instance, Pseudomonas strains have been isolated, which selectively deoxygenate the 1, 2 positions of substituted biarylethers and biarylketones.

- Microbial degradation of chloro, dichloro and trichloromethanes and carbonate chloride is also used to deal with the problem.

The different methods of effluent treatment processes are shown in figure:

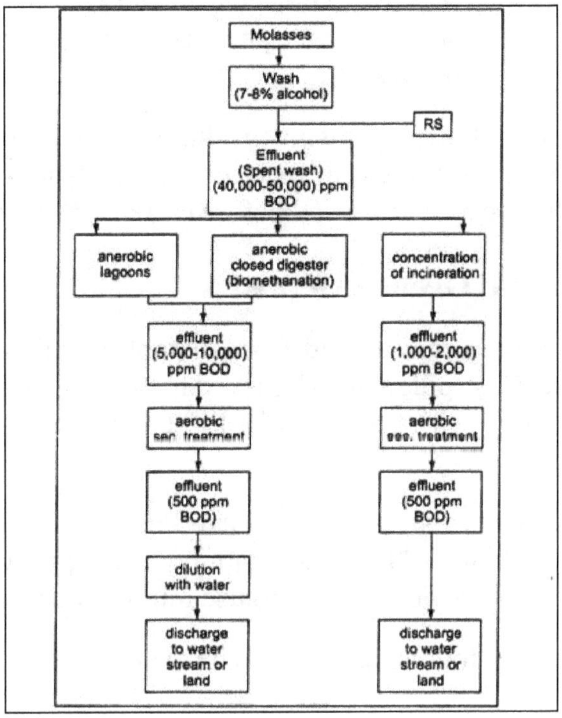

Different methods of effluent treatment.

Toxic Site Reclamation

Since the toxic constituents of industrial effluents are known in most cases, their treatment is easy. However, the waste disposal sites are more difficult to deal with, since their chemistry is often complex.

Although incineration (drying and then burning to ashes in a furnace) or chemical treatments are suggested for treatment of these sites, biotechnological approaches involving biodegradation are considered more attractive.

Companies like Bio Technical are working on treating polluted sites in situ. In this connection, it is interesting to note that the first patent ever awarded was given to Dr. Ananda Chakraborty for developing a microbe designed to biodegrade spilled oil. There has been a debate on whether or not to release genetically engineered microbes for treatment of toxic sites.

In view of the above, "Occidental Chemical Corporation" company of USA is conducting their work on effluents from toxic waste sites in closed reactors. One such experiment was conducted at Hyde Park landfill site at the north-west corner of Niagara, USA. This site occupies just over 6 hectares of land and contains an estimated 73,000 tonne of chemical waste.

The site was used from 1953 to 1975 to dispose- off a wide range of waste materials including chlorinated organic materials. For treatment of this site, a clay cover was placed on the landfill in 1978 and a leachate collection system was installed around the perimeter in 1979.

From this system, 230 m^3 of leachate per week was shipped to a nearby treatment facility in Niagara where the treatment involved carbon absorption, thus making the effluent suitable for discharge. This solution was not found suitable on a long term basis.

Therefore, the Hyde Park waste-water was subjected to bio treatment in sequencing batch reactors, where biodegradation rate of some more persistent chemicals was accelerated by adding bacterial strain isolated from the landfill site.

Further genetic manipulation of these organisms produced more than 100 recombinant bacteria, which have been tested for their usefulness in reactors. It was demonstrated that significant cost savings will be possible if SBR treatment is used before carbon absorption stage.

Removal of Spilled Oil and Grease Deposits

Oil spills from oil tankers on land surface and from distant oil spills have been recognised as a major environmental hazard. This spilled oil is believed to not only destroy the habitats of aquatic animals and fish, but also create health problems for local residents and causes long term damage to the environment.

Therefore, remediation efforts in this connection have been made, using chemical dispersants. These chemical dispersants are believed to cause major pollution problems in shallow water due to their toxic nature. Now attempts are being made to use microbiological degradation of oil.

In order to achieve this objective, following methods have been used:

Use of a Mixture of Bacterial Strains

A mixture of bacterial strain has also been used to clean oil-contaminated water reserves and water supplies. This could be illustrated using the following example. When 800,000 gallons of oily water once accumulated in the bottom of the ship "Queen Mary", there was a danger that this oily water will be discharged at the harbour causing death to marine life and spoiling beaches.

When a mixture of bacterial strains was introduced in the bottom of the ship, the oil was decomposed within 6 weeks. A mixture of bacterial strains is also being used for removal of grease deposited on the inner side of tanks in a ship. This technique may also be used for cleaning deposits of grease in pipes and vessels of a variety of industries.

Use of Oleophilic Fertilisers

In early years, application of oleophilic (oil-loving) fertilisers as food for oil-utilizing microbes was considered useful, since this would allow rapid growth and multiplication of indigenous microbes, thus speeding up the biodegradation process for removal of oil.

When 6,000 gallons of petrol leaked from a company in Pennsylvania in USA, it posed a serious health problem due to contamination of underground water supplies. Addition of oleophilic fertilisers on the sites expedited the biodegradation of petrol. In recent years, oil- utilised microbes—Pseudomonas aeruginosa—have been developed through genetic engineering.

Reducing Environmental Impact of Herbicides, Pesticides and Fertilisers

Majority of chemical herbicides, pesticides, and fertilizers that are used commercially cause environmental hazards.

Following solutions to this problem of pollution are being tried under biotechnology programmes:

Weed Control and Herbicides

Herbicides are being developed which will be environmentally safe. To allow the use of these herbicides for crop protection programmes, genetically engineered herbicide resistant plants have been produced in a number of crops.

These plants are being field-tested. Similarly, genetically engineered insect-resistant plants have also been successfully produced in some crops, thus raising the hope for reduced use of insecticides in future.

Pest Control and Bio Pesticides

Bacterial pesticides are also being developed. Heliothis complex, which lives in close association with plant roots, consists of 2 major crop pests— budworm and ball-worm.

Biological insecticides against both these insects are being prepared by transfer of a gene from Bacillus thuringiensis (Bt) into either a naturally occurring soil bacterium, or in a strain of Pseudomonas. Bt is already in the market since 1970 and in future these will be modified through techniques of genetic engineering and will be utilised against a variety of insects.

Several companies in USA—including Ecogen, Mycogen, Repligen and Zoe Con are involved in development and marketing of biological pesticides. Among these, Mycogen kills recombinant bacteria and applies them to the leaves of the crop plants.

Viral Pesticides

Have also been successfully developed and used for control of pest. These pesticides are environmentally safe and avoid risk of toxicity. These pesticides can also be used against the pest strains which have become resistant to chemical pesticides.

A number of entomopathogenic viruses belonging to the family Baculoviridae (or nuclear polyhedrosis viruses = NPV) and family Reoviridae (cytoplasmic polyhedrosis virus – CPV) have been used as safe and effective pesticides. These viruses kill specific pest species and have no adverse effect on useful insect pollinators, warm-blooded mammals or even man.

Some viral pesticides, with the name of pests they control are:

Class of virus	Pest	Trade name	Crops
1. N.P.V. Gypsy marth baculovirus	Tymantria clispar	Gypcheck	Hard wood trees
2. CPV CPV of Pine	Dendrotimus spectaabills	_	Pine, spruce etc.
3. Granulosis virus (GVS)	Cydia pomonetic	_	Apple and vegetables

Biofertiliser

To reduce the impact of excess chemical fertilizers in the field of agriculture, the bio-fertiliser is a potential tool.

Biosensors to detect Environmental Pollutants

Biosensors represent biophysical devices which will detect the presence and measure the quantities of specific substances in a variety of environments. These specific substances may include sugars, proteins or hormones in human body, pollutants in abiotic components of the environment including air/soil and water and a variety of toxins in the industrial effluents.

In designing a biosensor an enzyme or an antibody or even microbial cells are associated with microchip devices which are used for quantitative estimation of a substance.

A biosensor equipment consists of:

- A biologically sensing agent.
- A device for collecting the product obtained from the interaction between the substrate and the biosensor.
- A device for measuring thc quantity of this product, thus indirectly giving an estimate of the substrate.

The first reported application of biosensor dates back to 1860 when starch content was estimated with the help of malt extracts. Biosensor today is an analytical tool which consists of biological material in intimate contact with a transducer.

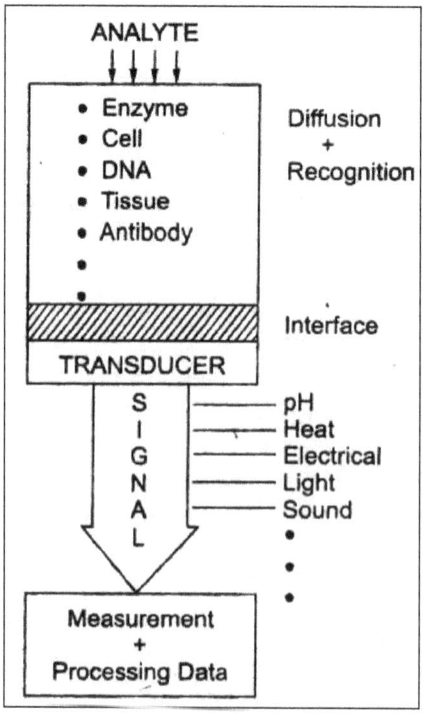

Schematics of biosensor.

It combines high selectivity of the biological/chemical material with high accuracy of solid state devices. However, interfacing poses difficult problems. The biological material can be membrane, enzyme, antibody/antigen, receptor, protein, intact cells, tissue or whole organ. It is necessary that the analyte should reach the reaction site in the biological material.

The reaction center can recognize the analyte and generate signals which can be measured by an appropriate transducer. The choice of transducer depends upon the signal generated by the biological material. The signal can be heat generated, light emitted/absorbed, pH change, electrons released/absorbed or chemical species produced.

The signal may or may not be amplified within a transducer. Weak signals require amplification prior to detection. Data processing can be done on a computer or recorded directly with a plotter. The first modern biosensor was assembled by Clark and Lyons, who modified oxygen electrode to sense glucose concentration. Since then a variety of biosensors have been developed.

Biosensors can be classified according to transducers employed.

The transducers are of different types:

- Electrochemical.
- Optical.
- Calorimetric.
- Field effect transducers.
- Piezoelectric.
- Microbial biosensor.
- Enzyme biosensor.

Electrochemical transducers (potentiometric or amperometric) are most popular because of their familiarity to chemists and biologists as well as simplicity of operation. It has added advantage that the same device can be used to sense different types of redox molecules, provided their redox potentials are distinctly different from one another.

In optical biosensors, light is focused on a membrane containing reagents which can combine with target molecules diffusing through membrane. As a result, reagents can absorb or fluorescence, which is, measured and thereby concentrations of target molecules are determined. There are two types of sensor materials used—bifurcated, and coated fiber devices.

Biosensors based on bio analytical calorimetry rest upon the fact that most of the biological reactions are exothermic. Most of the reactions are associated with high molar enthalpy changes (20-100 KJ per mol) in a single enzymatic step. An enzyme immobilized electrode coupled to calorimetric device forms thermometric sensor.

These are simple in operation, insensitive to optical properties of sample and possess high specificity. The most important aspect of this device is that it can be integrated with continuous flow systems and can easily detect 0.01 mM concentrations.

Semiconductor-based devices are ragged and can be mass-fabricated at low costs. The device consists of a chemically sensitive layer, integrated with solid state electronic circuits. It can be designed as a field effect transistor (FET) or in form of a capacitor. FET based device has built-in ampliation and associated electronic equipment is less complicated. It is immune to electromagnetic disturbances.

Conductometric or impedimetric measurements offer yet another possibility. The device can have either two terminal or four terminal measuring system. It is desirable that polarization of electrode and contamination of electrode surface due to oxide film formation should be minimum.

The electrodes should have large area and possess reversible electrochemistry. Au, Pt, Pd, Ti, Al, Fe electrodes are commonly used. These sensors are not popular to biologists and industries as the theory behind these measurements is unfamiliar to the researchers.

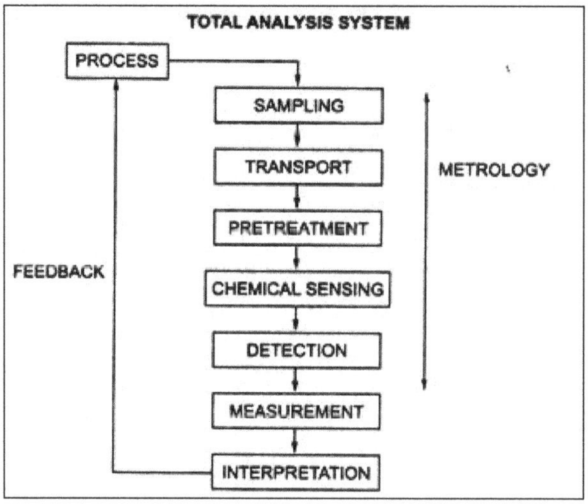

Biosensor in flow injection analysis system.

Microbial biosensors consist of a number of coupled or uncoupled enzymic steps. These are less sensitive to inhibitors and highly tolerant of variations in pH and temperature. These are cheaper and have long life times.

But there are several disadvantages viz.:

- Less selective and difficult to maintain.
- Response times are long and return to baseline at slower rate.
- Sensors are highly susceptible to contamination.

The immobilization of enzyme can be attained by physical entrapment, chemical attachment or by preparing thin solid films. Physical entrapment is a mild method and does not adversely affect catalytic activity of the enzyme.

This can be achieved by immobilizing enzyme-substrate complex as a whole. Chemical attachment can be of ionic or covalent linkage type. The covalent coupling should not involve amino acid residues which are essential to catalytic activity. Lastly, the use of biofilm for immobilization of enzyme is also very effective. Such enzyme immobilized film is packed in a dialysis bag for effective biosensor functioning.

ACTIVATED SLUDGE PROCESS

The activated sludge process is a type of wastewater treatment process for treating sewage or industrial wastewaters using aeration and a biological floc composed of bacteria and protozoa.

The general arrangement of an activated sludge process for removing carbonaceous pollution includes the following items: An aeration tank where air (or oxygen) is injected in the mixed liquor. This is followed by a settling tank (usually referred to as "final clarifier" or "secondary settling tank") to allow the biological flocs (the sludge blanket) to settle, thus separating the biological sludge from the clear treated water.

Activated sludge tank at Beckton sewage treatment plant - the white bubbles are due to the diffused air aeration system.

Purpose

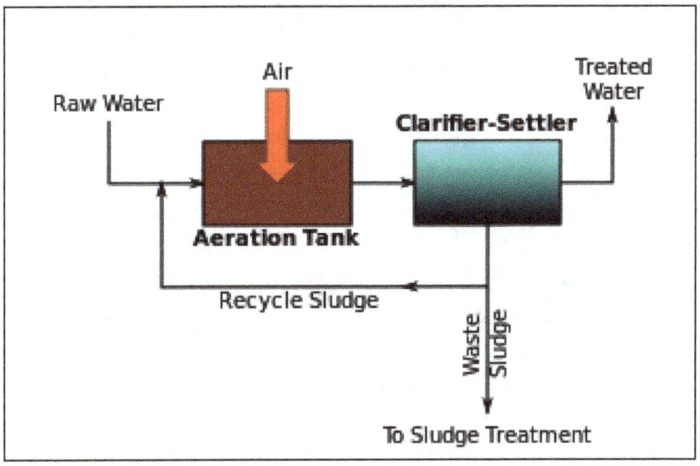

A generalized, schematic diagram of an activated sludge process.

Activated sludge addition (seeding) to a pilot scale membrane bioreactor.

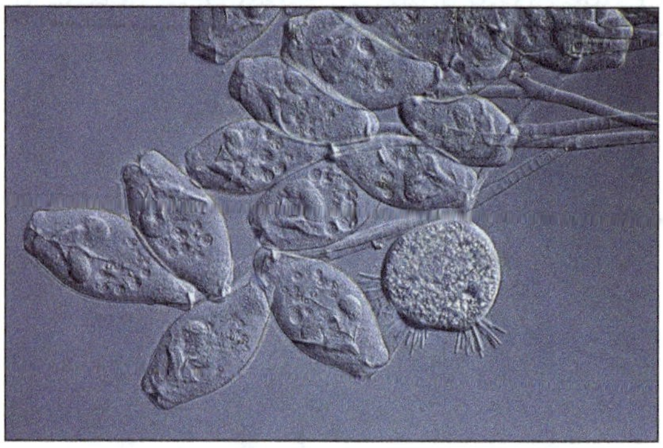

Activated sludge under the microscope.

In a sewage (or industrial wastewater) treatment plant, the activated sludge process is a biological process that can be used for one or several of the following purposes: Oxidizing carbonaceous biological matter, oxidizing nitrogenous matter: Mainly ammonium and nitrogen in biological matter, removing nutrients (nitrogen and phosphorus).

History

The Davyhulme Sewage Works Laboratory, where the activated sludge process was developed in the early 20th century.

The activated sludge process was discovered in 1913 in the United Kingdom by two engineers, Edward Ardern and W.T. Lockett, who were conducting research for the Manchester Corporation

Rivers Department at Davyhulme Sewage Works. This development led to arguably the single most significant improvement in public health and the environment during the course of the century.

In 1912, Dr. Gilbert Fowler, a scientist at the University of Manchester, observed experiments being conducted at the Lawrence Experiment Station at Massachusetts involving the aeration of sewage in a bottle that had been coated with algae. Fowler's engineering colleagues, Ardern and Lockett, experimented on treating sewage in a draw-and-fill reactor, which produced a highly treated effluent. They aerated the waste-water continuously for about a month and were able to achieve a complete nitrification of the sample material. Believing that the sludge had been activated (in a similar manner to activated carbon) the process was named activated sludge. Not until much later was it realized that what had actually occurred was a means to concentrate biological organisms, decoupling the liquid retention time (ideally, low, for a compact treatment system) from the solids retention time (ideally, fairly high, for an effluent low in BOD_5 and ammonia.)

Their results were published in their seminal 1914 paper, and the first full-scale continuous-flow system was installed at Worcester two years later. In the aftermath of the First World War the new treatment method spread rapidly, especially to the USA, Denmark, Germany and Canada. By the late 1930s, the activated sludge treatment became a well-known biological wastewater treatment process in those countries where sewer systems and sewage treatment plants were common.

Process Description

The process takes advantage of aerobic micro-organisms that can digest organic matter in sewage, and clump together (by flocculation) as they do so. It thereby produces a liquid that is relatively free from suspended solids and organic material, and flocculated particles that will readily settle out and can be removed.

The general arrangement of an activated sludge process for removing carbonaceous pollution includes the following items:

- Aeration tank where air (or oxygen) is injected in the mixed liquor.
- Settling tank (usually referred to as "final clarifier" or "secondary settling tank") to allow the biological flocs (the sludge blanket) to settle, thus separating the biological sludge from the clear treated water.

Treatment of nitrogenous matter or phosphate involves additional steps where the processes are managed to generate an anoxic zone so that phosphates can be solubilised in the reducing environment and oxides of nitrogen can be reduced to ammonium ion.

Bioreactor and Final Clarifier

The process involves air or oxygen being introduced into a mixture of screened, and primary treated sewage or industrial wastewater (wastewater) combined with organisms to develop a biological floc which reduces the organic content of the sewage. This material, which in healthy sludge is a brown floc, is largely composed of saprotrophic bacteria but also has an important protozoan flora component mainly composed of amoebae, Spirotrichs, Peritrichs including Vorticellids and a range of other filter-feeding species. Other important constituents include motile and sedentary

Rotifers. In poorly managed activated sludge, a range of mucilaginous filamentous bacteria can develop including Sphaerotilus natans which produces a sludge that is difficult to settle and can result in the sludge blanket decanting over the weirs in the settlement tank to severely contaminate the final effluent quality. This material is often described as sewage fungus but true fungal communities are relatively uncommon.

The combination of wastewater and biological mass is commonly known as mixed liquor. In all activated sludge plants, once the wastewater has received sufficient treatment, excess mixed liquor is discharged into settling tanks and the treated supernatant is run off to undergo further treatment before discharge. Part of the settled material, the sludge, is returned to the head of the aeration system to re-seed the new wastewater entering the tank. This fraction of the floc is called return activated sludge (R.A.S.).

The space required for a sewage treatment plant can be reduced by using a membrane bioreactor to remove some wastewater from the mixed liquor prior to treatment. This results in a more concentrated waste product that can then be treated using the activated sludge process.

Many sewage treatment plants use axial flow pumps to transfer nitrified mixed liquor from the aeration zone to the anoxic zone for denitrification. These pumps are often referred to as internal mixed liquor recycle pumps (IMLR pumps). The raw sewage, the RAS, and the nitrified mixed liquor are mixed by submersible mixers in the anoxic zones in order to achieve denitrification.

Sludge Production

Activated sludge is also the name given to the active biological material produced by activated sludge plants. Excess sludge is called "surplus activated sludge" or "waste activated sludge" and is removed from the treatment process to keep the ratio of biomass to food supplied in the wastewater in balance. This sewage sludge is usually mixed with primary sludge from the primary clarifiers and undergoes further sludge treatment for example by anaerobic digestion, followed by thickening, dewatering, composting and land application.

The amount of sewage sludge produced from the activated sludge process is directly proportional to the amount of wastewater treated. The total sludge production consists of the sum of primary sludge from the primary sedimentation tanks as well as waste activated sludge from the bioreactors. The activated sludge process produces about 70–100 kg/ML of waste activated sludge (that is kg of dry solids produced per ML of wastewater treated; one mega litre (ML) is 10^3 m^3). A value of 80 kg/ML is regarded as being typical. In addition, about 110–170 kg/ML of primary sludge is produced in the primary sedimentation tanks which most - but not all - of the activated sludge process configurations use.

Process Control

The general process control method is to monitor sludge blanket level, SVI (Sludge Volume Index), MCRT (Mean Cell Residence Time), F/M (Food to Microorganism), as well as the biota of the activated sludge and the major nutrients DO (Dissolved oxygen), nitrogen, phosphate, BOD (Biochemical oxygen demand), and COD (Chemical oxygen demand). In the reactor/aerator and clarifier system, the sludge blanket is measured from the bottom of the clarifier to the level of settled solids in the clarifier's water column; this, in large plants, can be done up to three times a day.

The SVI is the volume of settled sludge in millilitres occupied by 1g of dry sludge solids after 30 minutes of settling in a 1000 millilitre graduated cylinder. The MCRT is the total mass (lbs) of mixed liquor suspended solids in the aerator and clarifier divided by the mass flow rate (lbs/day) of mixed liquor suspended solids leaving as WAS and final effluent. The F/M is the ratio of food fed to the microorganisms each day to the mass of microorganisms held under aeration. Specifically, it is the amount of BOD fed to the aerator (lbs/day) divided by the amount (lbs) of Mixed Liquor Volatile Suspended Solids (MLVSS) under aeration. Some references use MLSS (Mixed Liquor Suspended Solids) for expedience, but MLVSS is considered more accurate for the measure of microorganisms. Again, due to expedience, COD is generally used, in lieu of BOD, as BOD takes five days for results.

Based on these control methods, the amount of settled solids in the mixed liquor can be varied by wasting activated sludge (WAS) or returning activated sludge (RAS).

There are a variety of types of activated sludge plants. These include:

Package Plants

There are a wide range of types of package plants, often serving small communities or industrial plants that may use hybrid treatment processes often involving the use of aerobic sludge to treat the incoming sewage. In such plants the primary settlement stage of treatment may be omitted. In these plants, a biotic floc is created which provides the required substrate. Package plants are designed and fabricated by specialty engineering firms in dimensions that allow for their transportation to the job site in public highways, typically width and height of 12 by 12 feet. Length varies with capacity with larger plants being fabricated in pieces and welded on site. Steel is preferred over synthetic materials (e.g., plastic) for its durability.

Package plants are commonly variants of extended aeration, to promote the 'fit and forget' approach required for small communities without dedicated operational staff. There are various standards to assist with their design.

To use less space, treat difficult waste, and intermittent flows, a number of designs of hybrid treatment plants have been produced. Such plants often combine at least two stages of the three main treatment stages into one combined stage. In the UK, where a large number of wastewater treatment plants serve small populations, package plants are a viable alternative to building a large structure for each process stage. In the US, package plants are typically used in rural areas, highway rest stops and trailer parks.

Oxidation Ditch

In some areas, where more land is available, sewage is treated in large round or oval ditches with one or more horizontal aerators typically called brush or disc aerators which drive the mixed liquor around the ditch and provide aeration. These are oxidation ditches, often referred to by manufacturer's trade names such as Pasveer, Orbal, or Carrousel. They have the advantage that they are relatively easy to maintain and are resilient to shock loads that often occur in smaller communities (i.e. at breakfast time and in the evening).

Oxidation ditches are installed commonly as 'fit and forget' technology, with typical design parameters of a hydraulic retention time of 24 – 48 hours, and a sludge age of 12 – 20 days. This compares with nitrifying activated sludge plants having a retention time of 8 hours, and a sludge age of 8 – 12 days.

Deep Shaft/Vertical Treatment

Where land is in short supply sewage may be treated by injection of oxygen into a pressured return sludge stream which is injected into the base of a deep columnar tank buried in the ground. Such shafts may be up to 100 metres deep and are filled with sewage liquor. As the sewage rises the oxygen forced into solution by the pressure at the base of the shaft breaks out as molecular oxygen providing a highly efficient source of oxygen for the activated sludge biota. The rising oxygen and injected return sludge provide the physical mechanism for mixing of the sewage and sludge. Mixed sludge and sewage is decanted at the surface and separated into supernatant and sludge components. The efficiency of deep shaft treatment can be high.

Surface aerators are commonly quoted as having an aeration efficiency of 0.5 - 1.5 kg O_2/kWh, diffused aeration as 1.5 - 2.5 kg O_2/KWh. Deep Shaft claims 5 – 8 kg O_2/kWh.

However, the costs of construction are high. Deep Shaft has seen the greatest uptake in Japan, because of the land area issues. Deep Shaft was developed by ICI, as a spin-off from their Pruteen process. In the UK it is found at three sites: Tilbury, Anglian water, treating a wastewater with a high industrial contribution; Southport, United Utilities, because of land space issues; and Billingham, ICI, again treating industrial effluent, and built (after the Tilbury shafts) by ICI to help the agent sell more.

DeepShaft is a patented, licensed, process. The licensee has changed several times and currently Noram Engineering sells it.

Surface-aerated Basins

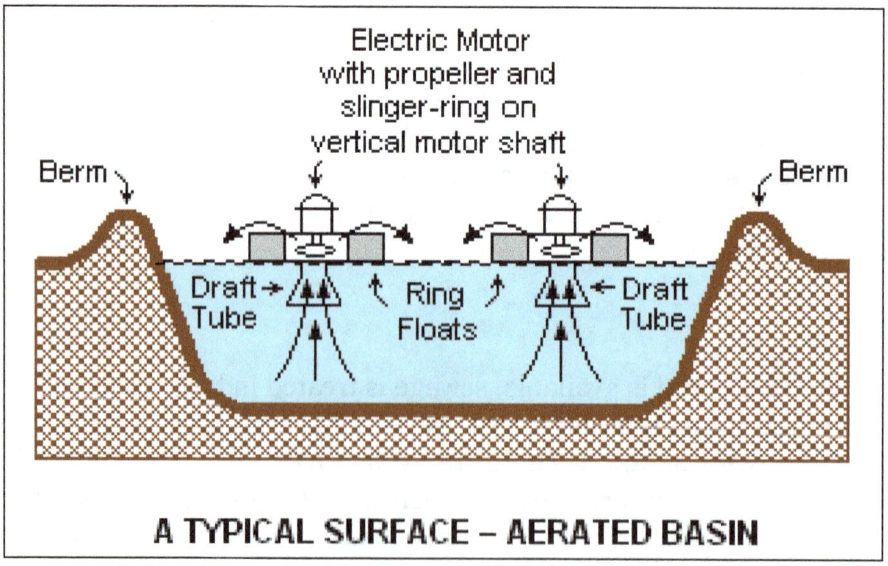

A Typical Surface-Aerated Basing (using motor-driven floating aerators).

Most biological oxidation processes for treating industrial wastewaters have in common the use of oxygen (or air) and microbial action. Surface-aerated basins achieve 80 to 90% removal of BOD with retention times of 1 to 10 days. The basins may range in depth from 1.5 to 5.0 metres and utilize motor-driven aerators floating on the surface of the wastewater.

In an aerated basin system, the aerators provide two functions: They transfer air into the basins required by the biological oxidation reactions, and they provide the mixing required for dispersing the air and for contacting the reactants (that is, oxygen, wastewater and microbes). Typically, the floating surface aerators are rated to deliver the amount of air equivalent to 1.8 to 2.7 kg O_2/kWh. However, they do not provide as good mixing as is normally achieved in activated sludge systems and therefore aerated basins do not achieve the same performance level as activated sludge units.

Biological oxidation processes are sensitive to temperature and, between 0 °C and 40 °C, the rate of biological reactions increase with temperature. Most surface aerated vessels operate at between 4 °C and 32 °C.

Sequencing Batch Reactors (SBRs)

Sequencing batch reactors (SBRs) treat wastewater in batches within the same vessel. This means that the bioreactor and final clarifier are not separated in space but in a timed sequence. The installation consists of at least two identically equipped tanks with a common inlet, which can be alternated between them. While one tank is in settle/decant mode the other is aerating and filling.

Aeration Methods

Diffused Aeration

Sewage liquor is run into deep tanks with diffuser grid aeration systems that are attached to the floor. These are like the diffused airstone used in tropical fish tanks but on a much larger scale. Air is pumped through the blocks and the curtain of bubbles formed both oxygenates the liquor and also provides the necessary mixing action. Where capacity is limited or the sewage is unusually strong or difficult to treat, oxygen may be used instead of air. Typically, the air is generated by some type of air blower.

Surface Aerators (Cones)

Vertically mounted tubes of up to 1-metre diameter extending from just above the base of a deep concrete tank to just below the surface of the sewage liquor. A typical shaft might be 10 metres high. At the surface end, the tube is formed into a cone with helical vanes attached to the inner surface. When the tube is rotated, the vanes spin liquor up and out of the cones drawing new sewage liquor from the base of the tank. In many works, each cone is located in a separate cell that can be isolated from the remaining cells if required for maintenance. Some works may have two cones to a cell and some large works may have 4 cones per cell.

Pure Oxygen Aeration

Pure oxygen activated sludge aeration systems are sealed-tank reactor vessels with surface aerator type impellers mounted within the tanks at the oxygen carbon liquor surface interface. The amount of oxygen entrainment, or DO (Dissolved Oxygen), can be controlled by a weir adjusted level control, and a vent gas oxygen controlled oxygen feed valve. Oxygen is generated on site by cryogenic distillation of air, pressure swing adsorption, or other methods. These systems are used where wastewater plant space is at a premium and high sewage throughput is required as high energy costs are involved in purifying oxygen.

Recent Developments

A new development of the activated sludge process is the Nereda process which produces a granular sludge that settles very well (the sludge volume index is reduced from 200-300 to 40 mL/g). A new process reactor system is created to take advantage of this quick settling sludge and is integrated into the aeration tank instead of having a separate unit outside. About 30 Nereda wastewater treatment plants worldwide are operational, under construction or under design, varying in size from 5,000 up to 858,000 person equivalent.

TRICKLING FILTER

A trickling filter is a type of wastewater treatment system. It consists of a fixed bed of rocks, coke, gravel, slag, polyurethane foam, sphagnum peat moss, ceramic, or plastic media over which sewage or other wastewater flows downward and causes a layer of microbial slime (biofilm) to grow, covering the bed of media. Aerobic conditions are maintained by splashing, diffusion, and either by forced-air flowing through the bed or natural convection of air if the filter medium is porous.

The terms trickle filter, trickling biofilter, biofilter, biological filter and biological trickling filter are often used to refer to a trickling filter. These systems have also been described as roughing filters, intermittent filters, packed media bed filters, alternative septic systems, percolating filters, attached growth processes, and fixed film processes.

A trickling filter plant: The effluent from the primary settling tanks is sprayed onto a bed of coarse gravel (Benfleet Sewage Treatment Plant).

Process Description

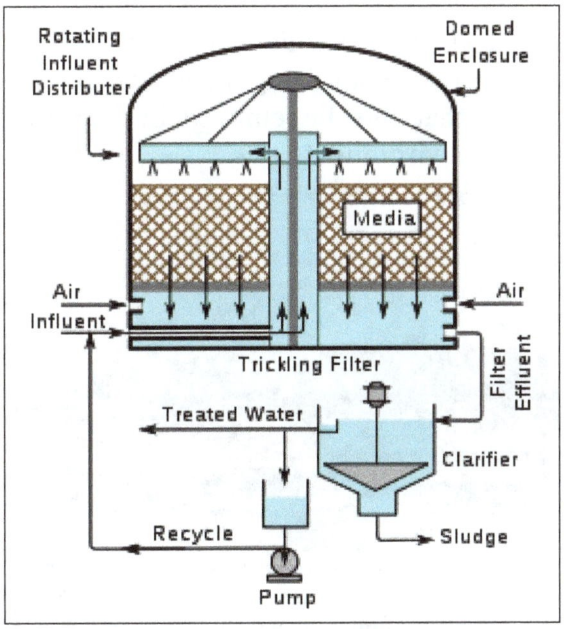

A typical complete trickling filter system.

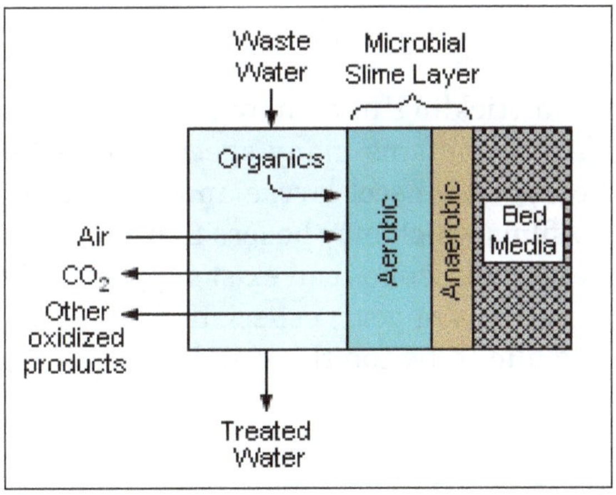

A schematic cross-section of the contact face of the bed of media in a trickling filter.

Typically, settled sewage flow enters at a high level and flows through the primary settlement tank. The supernatant from the tank flows into a dosing device, often a tipping bucket which delivers flow to the arms of the filter. The flush of water flows through the arms and exits through a series of holes pointing at an angle downwards. This propels the arms around distributing the liquid evenly over the surface of the filter media. Most are uncovered (unlike the accompanying diagram) and are freely ventilated to the atmosphere.

The removal of pollutants from the waste water stream involves both absorption and adsorption of organic compounds and some inorganic species such as nitrite and nitrate ions by the layer of microbial bio film. The filter media is typically chosen to provide a very high surface area to volume. Typical materials are often porous and have considerable internal surface area in addition to the external surface of the medium. Passage of the waste water over the media provides dissolved oxygen which the bio-film layer requires for the biochemical oxidation of the organic compounds and releases carbon dioxide gas, water and other oxidized end products. As the bio film layer thickens, it eventually sloughs off into the liquid flow and subsequently forms part of the secondary sludge. Typically, a trickling filter is followed by a clarifier or sedimentation tank for the separation and removal of the sloughed film. Other filters utilizing higher-density media such as sand, foam and peat moss do not produce a sludge that must be removed, but require forced air blowers and backwashing or an enclosed anaerobic environment.

Broken trickling filter unit at the sewage treatment plant in Norton, Zimbabwe, showing importance of maintenance to prevent structural failure

Biofilm

The bio-film that develops in a trickling filter may become several millimetres thick and is typically a gelatinous matrix that contains many species of bacteria, cilliates and amoeboid protozoa, annelids, round worms and insect larvae and many other micro fauna. This is very different from many other bio-films which may be less than 1 mm thick. Within the thickness of the biofilm both aerobic and anaerobic zones can exist supporting both oxidative and reductive biological processes. At certain times of year, especially in the spring, rapid growth of organisms in the film may cause the film to be too thick and it may slough off in patches leading to the "spring slough".

Design Considerations

A typical trickling filter is circular and between 10 metres and 20 metres across and between 2 metres to 3 metres deep. A circular wall, often of brick, contains a bed of filter media which in turn rests on a base of under-drains. These under-drains function both to remove liquid passing through the filter media but also to allow the free passage of air up through the filter media. Mounted in the center over the top of the filter media is a spindle supporting two or more horizontal perforated pipes which extend to the edge of the media. The perforations on the pipes are designed to allow an even flow of liquid over the whole area of the media and are also angled so that when liquid flows from the pipes the whole assembly rotates around the central spindle. Settled sewage is delivered to a reservoir at the centre of the spindle via some form of dosing mechanism, often a tipping bucket device on small filters.

Larger filters may be rectangular and the distribution arms may be driven by hydraulic or electrical systems.

Types

Single trickling filters may be used for the treatment of small residential septic tank discharges and very small rural sewage treatment systems. Larger centralized sewage treatment plants typically use many trickling filters in parallel.

Systems can be configured for single-pass use where the treated water is applied to the trickling filter once before being disposed of, or for multi-pass use where a portion of the treated water is cycled back and re-treated via a closed loop. Multi-pass systems result in higher treatment quality and assist in removing Total Nitrogen (TN) levels by promoting nitrification in the aerobic media bed and denitrification in the anaerobic septic tank. Some systems use the filters in two banks operated in series so that the wastewater has two passes through a filter with a sedimentation stage between the two passes. Every few days the filters are switched round to balance the load. This method of treatment can improve nitrification and de-nitrification since much of the carbonaceous oxidative material is removed on the first pass through the filters.

Media Types

Trickling may have a variety of types of filter media used to support the biofilm. Types of media most commonly used include coke, pumice, plastic matrix material, open-cell polyurethane foam, clinker, gravel, sand and geotextiles. Ideal filter medium optimizes surface area for microbial attachment, wastewater retention time, allows air flow, resists plugging, is mechanically robust in all weathers allowing walking access across the filter, and does not degrade. Some residential systems require forced aeration units which will increase maintenance and operational costs.

Industrial Wastewater Treatment

The treatment of industrial wastewater may involve specialized trickling filters which use plastic media and high flow rates. Wastewaters from a variety of industrial processes have been treated in trickling filters. Such industrial wastewater trickling filters consist of two types:

- Large tanks or concrete enclosures filled with plastic packing or other media.
- Vertical towers filled with plastic packing or other media.

The availability of inexpensive plastic tower packings has led to their use as trickling filter beds in tall towers, some as high as 20 meters. As early as the 1960s, such towers were in use at: The Great Northern Oil's Pine Bend Refinery in Minnesota; the Cities Service Oil Company Trafalgar Refinery in Oakville, Ontario and at a kraft paper mill.

The treated water effluent from industrial wastewater trickling filters is typically processed in a clarifier to remove the sludge that sloughs off the microbial slime layer attached to the trickling filter media as for other trickling filter applications.

Some of the latest trickle filter technology involves aerated biofilters of plastic media in vessels using blowers to inject air at the bottom of the vessels, with either downflow or upflow of the wastewater.

API OIL–WATER SEPARATOR

An API oil–water separator is a device designed to separate gross amounts of oil and suspended solids from the wastewater effluents of oil refineries, petrochemical plants, chemical plants, natural gas processing plants and other industrial oily water sources. The name is derived from the fact that such separators are designed according to standards published by the American Petroleum Institute (API).

Description of the Design and Operation

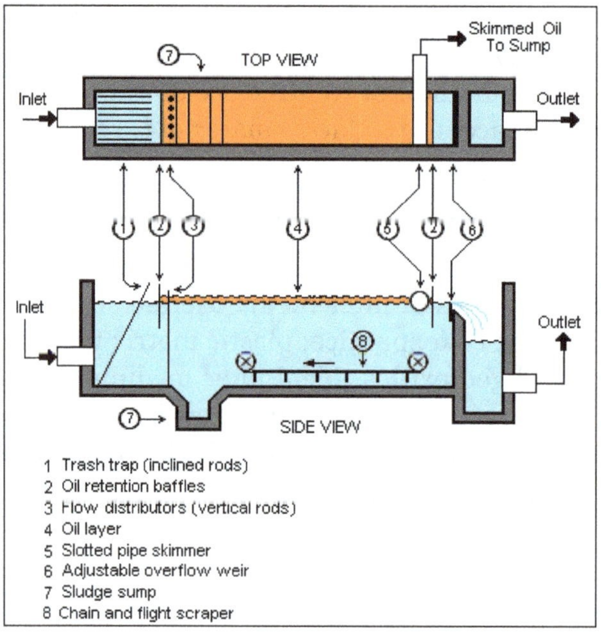

A typical gravimetric API separator.

The API separator is a gravity separation device designed using Stokes' law principles that define the rise velocity of oil droplets based on their density, size and water properties. The design of the separator is based on the specific gravity difference between the oil and the wastewater because that difference is much smaller than the specific gravity difference between the suspended solids and water. Based on that design criterion, most of the suspended solids will settle to the bottom of the separator as a sediment layer, the oil will rise to top of the separator, and the wastewater will be the middle layer between the oil on top and the solids on the bottom. The API Design Standards, when correctly applied, make adjustments to the geometry, design and size of the separator beyond simple Stokes Law principles. This includes allowances for water flow entrance and exit turbulence losses as well as other factors. API Specification 421 requires a minimum length to width ratio of 5:1 and minimum depth-to-width ratio of 0.3:0.5.

Typically in operation of API separators the oil layer, which may contain entrained water and attached suspended solids, is continually skimmed off. This removed oily layer may be re-processing to recover valuable products, or disposed of. The heavier bottom sediment layer is removed by a chain and flight scraper (or similar device) and a sludge pump.

Design Limitations

API design separators, and similar gravity tanks, are not intended to be effective when any of the following conditions apply to the feed conditions:

- Mean Oil droplets size in the feed is less than 150 micron.
- Oil density is greater than 925 kg/m3.
- Suspended solids are adhering to the oil meaning the 'effective' oil density is greater than 925 kg/m3.
- Water temperature less than 5 °C.
- There are high levels of dissolved hydrocarbons.

According to Stokes' Law, heavier oils require more retention time. In many cases where refineries have switched to heavier crude slates, the API separator's efficiency has declined.

Further Treatment of API Water Discharges

Because of performance limitations the water discharged from API type separators usually requires several further processing stages before the treated water can be discharged or reused. Further water treatment is designed to remove oil droplets smaller than 150 micron, dissolved materials and hydrocarbons, heavier oils or other contaminants not removed by the API. Secondary treatment technologies include dissolved air flotation (DAF), Anaerobic and Aerobic biological treatment, Parallel Plate Separators, Hydrocyclone, Walnut Shell Filters and Media filters.

Alternative Technologies

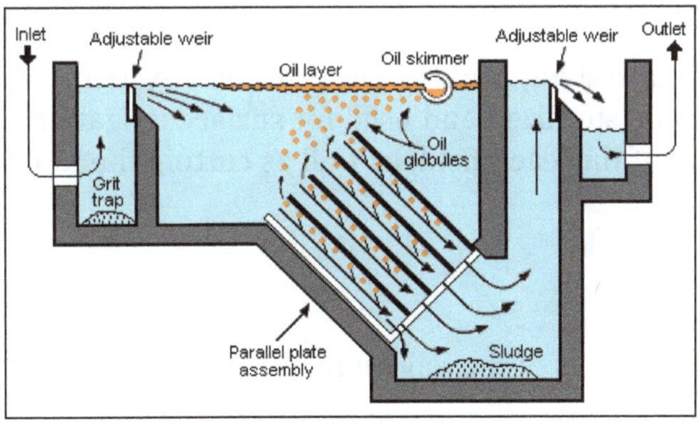

A typical parallel plate separator.

Plate separators, or Coalescing Plate Separators are similar to API separators, in that they are based on Stokes Law principles, but include inclined plate assemblies (also known as parallel packs). The underside of each parallel plate provides more surface for suspended oil droplets to coalesce into larger globules. Coalescing plate separators may not be effective in situation where water chemicals or suspended solids restrict or prevent oil droplets coalesce. In operation it is intended that sediment will slide down the topside of each parallel plate, however in many practical situations the sediment can adhere to the plates requiring periodic removal and cleaning. Such separators still depend upon the specific gravity between the suspended oil and the water.

However, the parallel plates can enhance the degree of oil-water separation for oil droplets above 50 micron in size. Alternatively parallel plate separators are added to the design of API Separators and require less space than a conventional API separator to achieve a similar degree of separation.

Other Oil–water Separation Applications

There are other applications requiring oil-water separation. For example:

- Oily water separators (OWS) for separating oil from the bilge water accumulated in ships as required by the international MARPOL Convention.
- Oil and water separators are commonly used in electrical substations. The transformers found in substations use a large amount of oil for cooling purposes. Moats are constructed surrounding unenclosed substations to catch any leaked oil, but these will also catch rainwater. Oil and water separators therefore provide a quicker and easier cleanup of an oil leak.

AEROBIC TREATMENT SYSTEM

An aerobic treatment system or ATS, often called (incorrectly) an aerobic septic system, is a small scale sewage treatment system similar to a septic tank system, but which uses an aerobic process for digestion rather than just the anaerobic process used in septic systems. These systems are commonly found in rural areas where public sewers are not available, and may be used for a single residence or for a small group of homes.

Unlike the traditional septic system, the aerobic treatment system produces a high quality secondary effluent, which can be sterilized and used for surface irrigation. This allows much greater flexibility in the placement of the leach field, as well as cutting the required size of the leach field by as much as half.

Process

The ATS process generally consists of the following phases:

- Pre-treatment stage to remove large solids and other undesirable substances.
- Aeration stage, where aerobic bacteria digest biological wastes.
- Settling stage allows undigested solids to settle. This forms a sludge that must be periodically removed from the system.
- Disinfecting stage, where chlorine or similar disinfectant is mixed with the water, to produce an antiseptic output. Another option is UV disinfection, where the water is exposed to UV light inside of a UV disinfection unit.

The disinfecting stage is optional, and is used where a sterile effluent is required, such as cases where the effluent is distributed above ground. The disinfectant typically used is tablets of calcium hypochlorite, which are specially made for waste treatment systems. The tablets are intended to break down quickly in sunlight. Stabilized forms of chlorine persist after the effluent is dispersed, and can kill plants in the leach field.

Since the ATS contains a living ecosystem of microbes to digest the waste products in the water, excessive amounts of items such as bleach or antibiotics can damage the ATS environment and reduce treatment effectiveness. Non-digestible items should also be avoided, as they will build up in the system and require more frequent sludge removal.

Types of Aerobic Treatment Systems

Small scale aerobic systems generally use one of two designs, fixed-film systems, or continuous flow, suspended growth aerobic systems (CFSGAS). The pre-treatment and effluent handling are similar for both types of systems, and the difference lies in the aeration stage.

Fixed Film Systems

Fixed film systems use a porous medium which provides a bed to support the biomass film that digests the waste material in the wastewater. Designs for fixed film systems vary widely, but fall into two basic categories (though some systems may combine both methods). The first is a system where the media is moved relative to the wastewater, alternately immersing the film and exposing it to air, while the second uses a stationary media, and varies the wastewater flow so the film is alternately submerged and exposed to air. In both cases, the biomass must be exposed to both wastewater and air for the aerobic digestion to occur. The film itself may be made of any suitable porous material, such as formed plastic or peat moss. Simple systems use stationary media, and rely on intermittent, gravity driven wastewater flow to provide periodic exposure to air and wastewater. A common moving media system is the rotating biological contactor (RBC), which uses disks rotating slowly on a horizontal shaft. Nearly 40 percent of the disks are submerged at any given time, and the shaft rotates at a rate of one or two revolutions per minute.

Continuous Flow and Suspended Growth Aerobic Systems

CFSGAS systems, as the name implies, are designed to handle continuous flow, and do not provide a bed for a bacterial film, relying rather on bacteria suspended in the wastewater. The suspension and aeration are typically provided by an air pump, which pumps air through the aeration chamber, providing a constant stirring of the wastewater in addition to the oxygenation. A medium to promote fixed film bacterial growth may be added to some systems designed to handle higher than normal levels of biomass in the wastewater.

Retrofit or Portable Aerobic Systems

Another increasingly common use of aerobic treatment is for the remediation of failing or failed anaerobic septic systems, by retrofitting an existing system with an aerobic feature. This class of product, known as aerobic remediation, is designed to remediate biologically failed and failing anaerobic distribution systems by significantly reducing the biochemical oxygen demand (BOD5) and total suspended solids (TSS) of the effluent. The reduction of the BOD5 and TSS reverses the developed bio-mat. Further, effluent with high dissolved oxygen and aerobic bacteria flow to the distribution component and digest the bio-mat.

Composting Toilets

Composting toilets are designed to treat only toilet waste, rather than general residential waste water, and are typically used with water-free toilets rather than the flush toilets associated with the above types of aerobic treatment systems. These systems treat the waste as a moist solid, rather than in liquid suspension, and therefore separate urine from feces during treatment to maintain the correct moisture content in the system. An example of a composting toilet is the clivus multrum (Latin for 'inclined chamber'), which consists of an inclined chamber that separates urine and feces and a fan to provide positive ventilation and prevent odors from escaping through the toilet. Within the chamber, the urine and feces are independently broken down not only by aerobic bacteria, but also by fungi, arthropods, and earthworms. Treatment times are very long, with a minimum time between removals of solid waste of a year; during treatment the volume of the solid waste is decreased by 90 percent, with most being converted into water vapor and carbon dioxide. Pathogens are eliminated from the waste by the long durations in inhospitable conditions in the treatment chamber.

Comparison to Traditional Septic Systems

The aeration stage and the disinfecting stage are the primary differences from a traditional septic system; in fact, an aerobic treatment system can be used as a secondary treatment for septic tank effluent. These stages increase the initial cost of the aerobic system, and also the maintenance requirements over the passive septic system. Unlike many other biofilters, aerobic treatment systems require a constant supply of electricity to drive the air pump increasing overall system costs. The disinfectant tablets must be periodically replaced, as well as the electrical components (air compressor) and mechanical components (air diffusers). On the positive side, an aerobic system produces a higher quality effluent than a septic tank, and thus the leach field can be smaller than that of a conventional septic system, and the output can be discharged in areas too environmentally sensitive for septic system output. Some aerobic systems recycle the effluent through a sprinkler system, using it to water the lawn where regulations approve.

Effluent Quality

Since the effluent from an ATS is often discharged onto the surface of the leach field, the quality is very important. A typical ATS will, when operating correctly, produce an effluent with less than 30 mg/liter BOD5, 25 mg/L TSS, and 10,000 cfu/mL fecal coliform bacteria. This is clean enough that it cannot support a biomat or "slime" layer like a septic tank.

ATS effluent is relatively odorless; a properly operating system will produce effluent that smells musty, but not like sewage. Aerobic treatment is so effective at reducing odors, that it is the preferred method for reducing odor from manure produced by farms.

ANAEROBIC BIOLOGICAL WASTEWATER TREATMENT

Anaerobic treatments on wastewater are normally implemented when treating more concentrated wastewater. The anaerobic sludge contains various groups of micro organisms that work together to eventually convert organic material to biogas via hydrolysis and acidification. Biogas typically consists of 70% methane (CH_4) and 30% carbon dioxide (CO_2) with residual fractions of other gases (e.g. H_2 and H_2S). The methane can be used as an energy source. Anaerobic reactors can be implemented in a variety of ways. The figure shows a contact reactor and an upflow reactor.

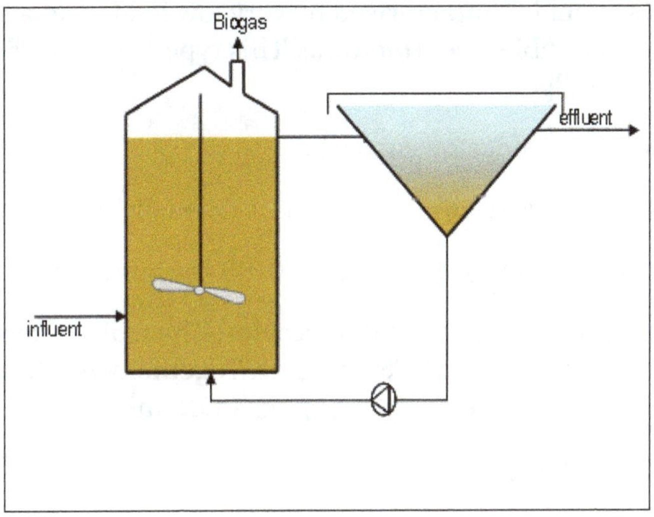

Anaerobic contact reactor.

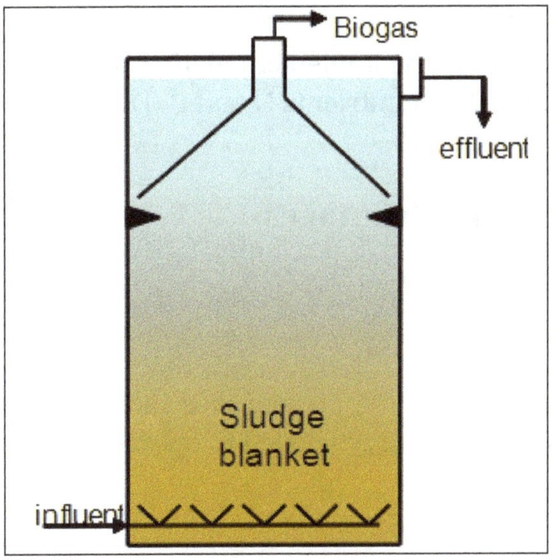

Upflow reactor.

The contact reactor is comparable with a conventional active sludge system, but under anaerobic conditions. The sludge is mixed with wastewater in the reactor and is then separated in the sedimentation tank and returned to the reactor.

In the anaerobic upflow reactor, the influent is introduced at the bottom of the vertical reactor. The sludge in the reactor is primarily grain shaped and forms a blanket in the reactor, with the most compact sludge grains at the bottom and the lighter grains and heavier sludge floccules above it. Very light sludge floccules will be released by the upward flow, but can potentially be collected in a sedimentation tank. The biogas is collected and disposed of at the top of the reactor, separately from the partly purified water and the sludge.

In addition to the contact reactor and the upflow reactor, other types are also available:

- Conventional digester.

  This type is primarily implemented for the fermentation of RWZI sludge and liquid organic waste. The system is characterised by very low loads and a large volume in order to achieve the longest possible retention time. This type of reactor does not include recirculation of anaerobic sludge.

- Packed anaerobic filter (sludge on carrier).

  This reactor is filled with carrier material and is normally used as an upflow reactor.

- UASB (upflow anaerobic sludge blanket) or EGSB (expanded granular sludge bed).

  Both systems are variations of the upflow reactor. The main difference between the two is the increased recirculation of the EGSB reactor. Together with the prominent sludge grain, this enables higher loads in the EGSB (15-30 kg COD/m^3/day).

- Anaerobic membrane reactor.

  This type of application uses membranes for sludge-water separation. To date, little use has been made of this system.

An extra purification phase will often be implemented after anaerobic purification, e.g. for the removal of residual fractions of COD and nutrients N and P. This often involves the use of an aerobic post-purification treatment.

Specific Advantages and Disadvantages

Advantages:

- Formation of biogas.

  The encountered organic pollutant is converted into biogas with a high energetic value. This, for example, allows the energy needed to operate the water purification system to be fully or partly recuperated.

- High loads.

  The volumetric load (COD load per m^3 active volume per day) in an anaerobic reactor is typically 5 to 10 times higher than aerobic wastewater purification.

- Very little sludge production.

Sludge growth in an anaerobic reactor is 4-5 times lower than in an aerobic system.

Working in campaigns is possible. If the anaerobic sludge is not fed, it will hibernate, which means longer periods without food can be spanned without excessive sludge mortality. The system will almost immediately become active after re-start.

Disadvantages:

- Incomplete break-down of organic compounds: Need for post-purification via, for example, aerobic purification.
- No thorough nutrient removal: Again, later aerobic purification with nutrient removal is often needed.
- Most efficient purification in the mesophilic range, i.e. between 30-37°C, whereby the influent must be heated in most cases.
- Less robust system with regards to toxicity and inhibition.
- Risk of odour problems.

Application

Anaerobic purification is implemented in various sectors. In the food sector, this technique is regularly used to reduce the high cost of aerobic waste purification by partially breaking down the organic load and converting into biogas.

Anaerobic processes are also frequently used to ferment aerobic sludge and fluid organic waste.

Boundary Conditions

Anaerobic purification is often preceded by a tank for buffering and conditioning the wastewater. The typical retention time is two days. Acidification takes place up to pH 5.5 or 6, along with hydrolysis of the suspended matter.

Anaerobic reactors are typically used for wastewater purification when the wastewater complies with the following conditions:

- Average to high COD concentration.
- Temperatures from approximately 20°C.
- Average to low salt concentrations.
- Low sulphate concentration (ratio of COD concentration over S04 concentration less than 10).
- Low concentration of fats/oils.
- Absence of toxic components.

The effluent from anaerobic purification will normally not comply with applicable discharge norms and will require an extra purification step. Anaerobic post-purification will be selected in many cases, whereby the organic and nutrient load will be reduced.

Effectiveness

The anaerobic reactor can be implemented for removing the following parameters:

- COD: On average, the reactor will remove 80-90% of ingoing COD.
- N: Is incorporated into the sludge at a rate of 13g N per 1000g removed COD.
- P: Is incorporated into the sludge at a rate of 3g P per 1000g removed COD.

ZERO-SLUDGE TREATMENT OF INDUSTRIAL WASTEWATER

Industrial wastewater treatment is an unusual engineering challenge. Where else in industry are the designers asked to provide a process that will produce a very high level of efficiency and reliability while being given an unpredictable, inconsistent and highly variable raw material with which to work? This conundrum is further compounded if, as is often the case, the waste is highly concentrated, high in salts or hot.

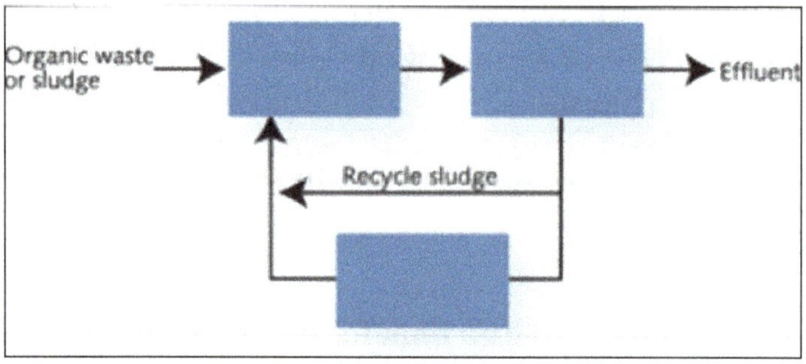

Schematic of the AFC^{sm} Process.

Industrial wastewater treatment is an unusual engineering challenge. Where else in industry are the designers asked to provide a process that will produce a very high level of efficiency and reliability while being given an unpredictable, inconsistent and highly variable raw material with which to work? This conundrum is further compounded if, as is often the case, the waste is highly concentrated, high in salts or hot.

High concentrations of organics (>5,000 mg/LBOD) pose two challenges. The mere concentration issue means even very efficient treatment can leave a significant effluent concentration (98% efficient treatment of a stream with 10,000 mg/L BOD still leaves 200 mg/L in the effluent). Further, a stream with a high concentration of BOD will release a lot of heat from bio-oxidation of organics, which can cause mixed liquor temperature to exceed the mesophilic range; this typically results in deflocculation, foaming and high effluent TSS.

High salt concentration (>1,000 mg/L) contributes to high effluent TSS as dispersed solids increase with ionic strength of the solution.

Hot wastewater typically must be cooled prior to treatment to insure the mixed liquor stays in the normal mesophilic temperature range (18-35°C), where most plants operate, and foaming and effluent clarity are easier to control. Biological activity is much faster and more efficient at higher temperatures, however, and thus much smaller reactors are required.

When the desired level of clean effluent finally is achieved, the owner is faced with the vexing question: "How do I pay for the river of noxious waste sludge we're generating every day?"

And at typical sludge generation rates (0.5-0.9 gTSS/gBOD), there are literally millions of pounds of sludge generated daily in the United States alone. Its fate will be the question of the decade for industrial wastewater treatment plant owners and operators. Europe has already passed severe restrictions on disposal of sludge in landfills, with total bans in sight. California has followed suit with rigorous new legislation proposed. This trend will spread as nobody wants industrial sludge hauled into his or her backyard. Many state regulatory "sludge gurus" have said the cost of sludge treatment is going to spiral upwards quickly and permanently.

But there are alternatives to the effort and expense of endlessly hauling sludge from your plant.

No-sludge Solution

The AFCsm, is a catalytically enhanced membrane bioreactor (MBR). Raw influent is fed to a self-heating thermophilic bioreactor for treatment. Mixed liquor is recirculated through an ultrafiltration (UF) membrane system for solids separation (which eliminates flocculation and settling problems caused by high salt and temperature). A small portion of this flow goes through a small chemical treatment reactor. The chemical treatment step partially solubilizes the excess biosolids before they're returned to the bioreactor for ultimate digestion and destruction. Field experience has shown the chemical treatment step also acts to suppress foaming in thermophilic reactors and enhances membrane flux rates.

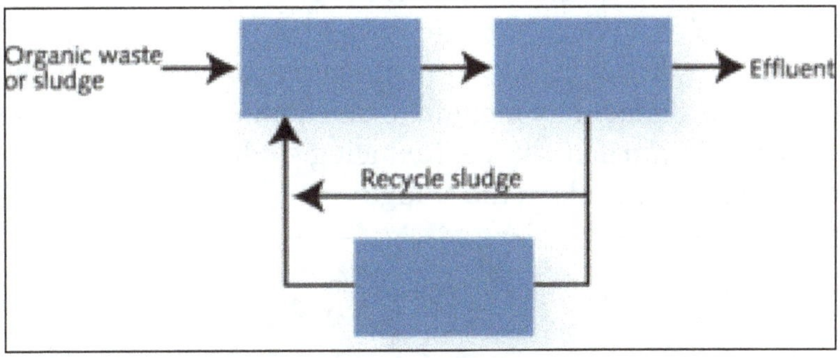

Schematic of the AFCsm Process.

Characteristics of catalytically enhanced MBR systems include:

- Little or no waste sludge production: Most full-scale systems produce zero waste biosolids.
- Handles hot wastes: Readily operates from 10-70°C.
- Very small footprint: Requires only 20% of the footprint compared to conventional activated sludge systems. High temperature operation speeds up kinetics 100-300% over conventional systems; membrane system keeps solids in - with MLSS levels up to 30,000 mg/L typical.

- Simple Operation: Operation is simplified since the catalyzed MBR largely frees the operator from worry over mixed liquor flocculation and settling and clarifier solids management, etc. The UF membrane insures effluent with less than 5 mg/L TSS. Finally, there's little or no excess sludge to waste-digest-stabilize-thicken-dewater-store-transport-dispose of (and thus no associated equipment to build and operate).

SEWAGE SLUDGE TREATMENT

Sewage sludge treatment describes the processes used to manage and dispose of sewage sludge produced during sewage treatment. Sludge is mostly water with lesser amounts of solid material removed from liquid sewage. Primary sludge includes settleable solids removed during primary treatment in primary clarifiers. Secondary sludge separated in secondary clarifiers includes treated sewage sludge from secondary treatment bioreactors.

Sludge treatment is focused on reducing sludge weight and volume to reduce disposal costs, and on reducing potential health risks of disposal options. Water removal is the primary means of weight and volume reduction, while pathogen destruction is frequently accomplished through heating during thermophilic digestion, composting, or incineration. The choice of a sludge treatment method depends on the volume of sludge generated, and comparison of treatment costs required for available disposal options. Air-drying and composting may be attractive to rural communities, while limited land availability may make aerobic digestion and mechanical dewatering preferable for cities, and economies of scale may encourage energy recovery alternatives in metropolitan areas.

Energy may be recovered from sludge through methane gas production during anaerobic digestion or through incineration of dried sludge, but energy yield is often insufficient to evaporate sludge water content or to power blowers, pumps, or centrifuges required for dewatering. Coarse primary solids and secondary sewage sludge may include toxic chemicals removed from liquid sewage by sorption onto solid particles in clarifier sludge. Reducing sludge volume may increase the concentration of some of these toxic chemicals in the sludge.

Dried, anaerobically digested sludge.

Treatment Processes

Thickening

A sewage sludge thickener.

Thickening is often the first step in a sludge treatment process. Sludge from primary or secondary clarifiers may be stirred (often after addition of clarifying agents) to form larger, more rapidly settling aggregates. Primary sludge may be thickened to about 8 or 10 percent solids, while secondary sludge may be thickened to about 4 percent solids. Thickeners often resemble a clarifier with the addition of a stirring mechanism. Thickened sludge with less than ten percent solids may receive additional sludge treatment while liquid thickener overflow is returned to the sewage treatment process.

Dewatering

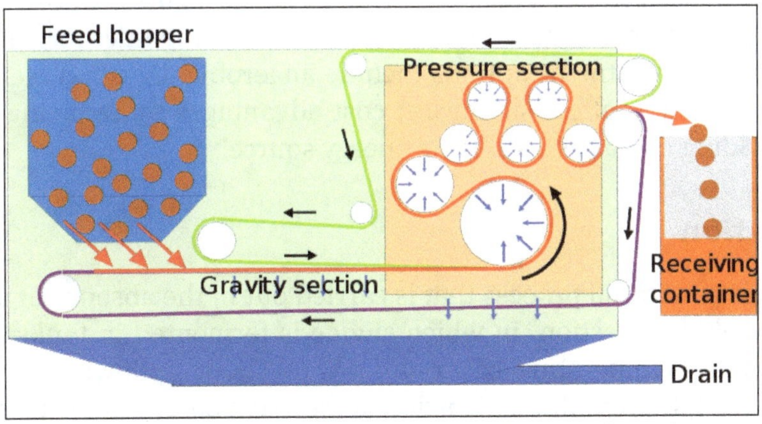

Schematic of a belt filter press to dewater sewage sludge. Filtrate is extracted initially by gravity, then by squeezing the cloth through rollers.

Water content of sludge may be reduced by centrifugation, filtration, and evaporation to reduce transportation costs of disposal, or to improve suitability for composting. Centrifugation may be a preliminary step to reduce sludge volume for subsequent filtration or evaporation. Filtration may occur through underdrains in a sand drying bed or as a separate mechanical process in a belt filter press. Filtrate and centrate are typically returned to the sewage treatment process. After dewatering sludge may be handled as a solid containing 50 to 75 percent water. Dewatered sludges with higher moisture content are usually handled as liquids.

Sidestream Treatment Technologies

Sludge treatment technologies that are used for thickening or dewatering of sludge have two products: the thickened or dewatered sludge, and a liquid fraction which is called sludge treatment liquids, sludge dewatering streams, liquors, centrate (if it stems from a centrifuge), filtrate (if it stems from a belt filter press) or similar. This liquid requires further treatment as it is high in nitrogen and phosphorus, particularly if the sludge has been anaerobically digested. The treatment can take place in the sewage treatment plant itself (by recycling the liquid to the start of the treatment process) or as a separate process.

Phosphorus Recovery

One method for treating sludge dewatering streams is by using a process that is also used for phosphorus recovery. Another benefit for sewage treatment plant operators of treating sludge dewatering streams for phosphorus recovery is that it reduces the formation of obstructive struvite scale in pipes, pumps and valves. Such obstructions can be a maintenance headache particularly for biological nutrient removal plants where the phosphorus content in the sewage sludge is elevated. For example, the Canadian company Ostara Nutrient Recovery Technologies is marketing a process based on controlled chemical precipitation of phosphorus in a fluidized bed reactor that recovers struvite in the form of crystalline pellets from sludge dewatering streams. The resulting crystalline product is sold to the agriculture, turf and ornamental plants sectors as fertiliser under the registered trade name "Crystal Green".

Digestion

Many sludges are treated using a variety of digestion techniques, the purpose of which is to reduce the amount of organic matter and the number of disease-causing microorganisms present in the solids. The most common treatment options include anaerobic digestion, aerobic digestion, and composting. Sludge digestion offers significant cost advantages by reducing sludge quantity by nearly 50% and providing biogas as a valuable energy source.

Anaerobic Digestion

Anaerobic digestion is a bacterial process that is carried out in the absence of oxygen. The process can either be thermophilic digestion, in which sludge is fermented in tanks at a temperature of 55 °C, or mesophilic, at a temperature of around 36 °C. Though allowing shorter retention time (and thus smaller tanks), thermophilic digestion is more expensive in terms of energy consumption for heating the sludge.

Mesophilic anaerobic digestion (MAD) is also a common method for treating sludge produced at sewage treatment plants. The sludge is fed into large tanks and held for a minimum of 12 days to allow the digestion process to perform the four stages necessary to digest the sludge. These are hydrolysis, acidogenesis, acetogenesis, and methanogenesis. In this process the complex proteins and sugars are broken down to form more simple compounds such as water, carbon dioxide, and methane.

Anaerobic digestion generates biogas with a high proportion of methane that may be used to both heat the tank and run engines or microturbines for other on-site processes. Methane generation is a key advantage of the anaerobic process. Its key disadvantage is the long time required for the process (up to 30 days) and the high capital cost. Many larger sites utilize the biogas for combined heat and power, using the cooling water from the generators to maintain the temperature of the digestion plant at the required 35 ± 3 °C. Sufficient energy can be generated in this way to produce more electricity than the machines require.

Aerobic Digestion

Aerobic digestion is a bacterial process occurring in the presence of oxygen resembling a continuation of the activated sludge process. Under aerobic conditions, bacteria rapidly consume organic matter and convert it into carbon dioxide. Once there is a lack of organic matter, bacteria die and are used as food by other bacteria. This stage of the process is known as *endogenous respiration*. Solids reduction occurs in this phase. Because the aerobic digestion occurs much faster than anaerobic digestion, the capital costs of aerobic digestion are lower. However, the operating costs are characteristically much greater for aerobic digestion because of energy used by the blowers, pumps and motors needed to add oxygen to the process. However, recent technological advances include non-electric aerated filter systems that use natural air currents for the aeration instead of electrically operated machinery.

Aerobic digestion can also be achieved by using diffuser systems or jet aerators to oxidize the sludge. Fine bubble diffusers are typically the more cost-efficient diffusion method, however, plugging is typically a problem due to sediment settling into the smaller air holes. Coarse bubble diffusers are more commonly used in activated sludge tanks or in the flocculation stages. A key component for selecting diffuser type is to ensure it will produce the required oxygen transfer rate.

Composting

Composting is an aerobic process of mixing sewage sludge with agricultural byproduct sources of carbon such as sawdust, straw or wood chips. In the presence of oxygen, bacteria digesting both the sewage sludge and the plant material generate heat to kill disease-causing microorganisms and parasites. Maintenance of aerobic conditions with 10 to 15 percent oxygen requires bulking agents allowing air to circulate through the fine sludge solids. Stiff materials like corn cobs, nut shells, shredded tree-pruning waste, or bark from lumber or paper mills better separate sludge for ventilation than softer leaves and lawn clippings. Light, biologically inert bulking agents like shredded tires may be used to provide structure where small, soft plant materials are the major source of carbon.

Uniform distribution of pathogen-killing temperatures may be aided by placing an insulating blanket of previously composted sludge over aerated composting piles. Initial moisture content of the composting mixture should be about 50 percent; but temperatures may be inadequate for pathogen reduction where wet sludge or precipitation raises compost moisture content above 60 percent. Composting mixtures may be piled on concrete pads with built-in air ducts to be covered by a layer of unmixed bulking agents. Odors may be minimized by using an aerating blower drawing vacuum through the composting pile via the underlying ducts and exhausting through a filtering pile of previously composted sludge to be replaced when moisture content reaches 70 percent. Liquid accumulating in the underdrain ducting may be returned to the sewage treatment plant; and composting pads may be roofed to provide better moisture content control.

After a composting interval sufficient for pathogen reduction, composted piles may be screened to recover undigested bulking agents for re-use; and composted solids passing through the screen may be used as a soil amendment material with similar benefits to peat. The optimum initial carbon-to-nitrogen ratio of a composting mixture is between 26-30:1; but the composting ratio of agricultural byproducts may be determined by the amount required to dilute concentrations of toxic chemicals in the sludge to acceptable levels for the intended compost use. Although toxicity is low in most agricultural byproducts, suburban grass clippings may have residual herbicide levels detrimental to some agricultural uses; and freshly composted wood byproducts may contain phytotoxins inhibiting germination of seedlings until detoxified by soil fungi.

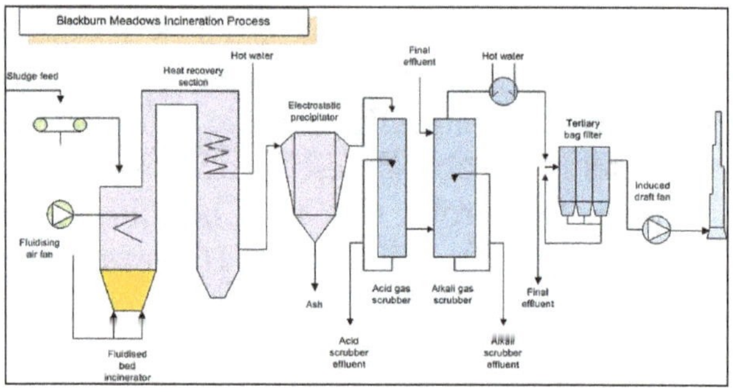

Sludge incineration process schematic.

Sewage sludge after drying in a sludge drying bed.

Incineration

Incineration of sludge is less common because of air emissions concerns and the supplemental fuel (typically natural gas or fuel oil) required to burn the low calorific value sludge and vaporize residual water. On a dry solids basis, the fuel value of sludge varies from about 9,500 British thermal units per pound (980 cal/g) of undigested sewage sludge to 2,500 British thermal units per pound (260 cal/g) of digested primary sludge. Stepped multiple hearth incinerators with high residence time and fluidized bed incinerators are the most common systems used to combust wastewater sludge. Co-firing in municipal waste-to-energy plants is occasionally done, this option being less expensive assuming the facilities already exist for solid waste and there is no need for auxiliary fuel. Incineration tends to maximize heavy metal concentrations in the remaining solid ash requiring disposal; but the option of returning wet scrubber effluent to the sewage treatment process may reduce air emissions by increasing concentrations of dissolved salts in sewage treatment plant effluent.

This simple evaporative sludge drying bed illustrates the initial consistency of primary sludge being discharged from the primary settling tank via the pipe in the foreground.

Drying Beds

Simple sludge drying beds are used in many countries, particularly in developing countries, as they are a cheap and simple method to dry sewage sludge. Drainage water must be captured; drying beds are sometimes covered but usually left uncovered. Mechanical devices to turn over the sludge in the initial stages of the drying process are also available on the market.

Drying beds are typically composed of four layers consisting of gravel and sand. The first layer is coarse gravel that is 15 to 20 centimeters thick. Followed by fine gravel that is 10 centimeters thick. The third layer is sand that can be between 10 and 15 centimeters and serves as the filter between the sludge and gravel. Sludge dries up and water percolates to the first layer that is collected at the drainage pipe that is beneath all layers.

Emerging Technologies

The thermal hydrolysis system at the Blue Plains treatment plant.

- Phosphorus recovery from sewage sludge or from sludge dewatering streams is receiving increased attention particularly in Sweden, Germany and Canada, as phosphorus is a limited resource (a concept also known as "peak phosphorus") and is needed as fertilizer to feed a growing world population. Phosphorus recovery methods from wastewater or sludge can be categorized by the origin of the used matter (wastewater, sludge liquor, digested or non-digested sludge, ash) or by the type of recovery processes (precipitation, wet-chemical extraction and precipitation, thermal treatment). Research on phosphorus recovery methods from sewage sludge has been carried out in Sweden and Germany since around 2003, but the technologies currently under development are not yet cost effective, given the current price of phosphorus on the world market.

- The Omni Processor is a process that is currently under development that treats sewage sludge and can generate a surplus of electrical energy if the input materials have the right level of dryness.
- Thermal depolymerization produces light hydrocarbons from sludge heated to 250 °C and compressed to 40 MPa.
- Thermal hydrolysis is a two-stage process combining high-pressure boiling of sludge, followed by a rapid decompression. This combined action sterilizes the sludge and makes it more biodegradable, which improves digestion performance. Sterilization destroys pathogens in the sludge resulting in it exceeding the stringent requirements for land application (agriculture). Thermal hydrolysis systems are operating at sewage treatment plants in Europe, China and North America, and can generate electricity as well as high quality sludge.

Disposal or use as Fertilizer

When a liquid sludge is produced, further treatment may be required to make it suitable for final disposal. Sludges are typically thickened and dewatered to reduce the volumes transported off-site for disposal. Processes for reducing water content include lagooning in drying beds to produce a cake that can be applied to land or incinerated; pressing, where sludge is mechanically filtered, often through cloth screens to produce a firm cake; and centrifugation where the sludge is thickened by centrifugally separating the solid and liquid. Sludges can be disposed of by liquid injection to land or by disposal in a landfill.

There is no process which completely eliminates the need to dispose of treated sewage sludge.

Much sludge originating from commercial or industrial areas is contaminated with toxic materials that are released into the sewers from the industrial processes. Elevated concentrations of such materials may make the sludge unsuitable for agricultural use and it may then have to be incinerated or disposed of to landfill.

Examples: The Edmonton Composting Facility, in Edmonton, Alberta, Canada, is the largest sewage sludge composting site in North America.

Sewage sludge can be superheated and converted it into pelletized granules that are high in nitrogen and other organic materials. In New York City, for example, several sewage treatment plants have dewatering facilities that use large centrifuges along with the addition of chemicals such as polymer to further remove liquid from the sludge. The product which is left is called "cake," and that is picked up by companies which turn it into fertilizer pellets. This product, also called biosolid, is then sold to local farmers and turf farms as a soil amendment or fertilizer, reducing the amount of space required to dispose of sludge in landfills.

In the very large metropolitan areas of southern California inland communities return sewage sludge to the sewer system of communities at lower elevations to be reprocessed at a few very large treatment plants on the Pacific coast. This reduces the required size of interceptor sewers and allows local recycling of treated wastewater while retaining the economy of a single sludge processing facility and is an example of how sewage sludge can help solve an energy crisis.

BRINE WASTE TREATMENT

Brine waste treatment is a set of one or more processes used to separate out contaminants and adjust the salinity of a brine effluent stream. Waste brines can arise from use as process liquids or ingredients in manufacturing, from use as regenerants for ion exchange (IX) resins, and as byproducts of various industrial processes, including desalination, boiler and cooling tower feedwater treatment, reverse osmosis (RO), energy recovery such as fracking and acid mine drainage, among others. Because brines are used in such a wide variety of applications, waste brine treatment strategies may draw on a number of different technologies and processes.

Why Treat Brine Waste Prior to Disposal?

Brine effluent will typically require some form of treatment prior to disposal, however, the specific type of treatment required will depend upon a few factors, including:

- Stream content and contaminants.
- The chosen method of disposal.
- Relevant environmental and discharge regulations, standards, and limitations.

An ideal brine waste disposal strategy will balance the costs associated with each of these factors.

Brine Waste Stream Contents

All waste brines share a high salinity (or salt concentration), but streams might vary significantly from one to the next in terms of other constituents present. This is because various materials can end up in the brine stream in the course of normal operations.

Common contaminants found in brine waste include:

- Suspended solids and particles.
- Antiscaling agents.
- Heavy metals.
- Microorganisms.
- Organic material .
- Oil and grease.

Contaminants such as these pose a threat to human health and can also damage ecosystems at the discharge site. If toxic or hazardous materials are present in your brine stream, chances are that some form of brine waste treatment will need to be implemented.

Brine Waste Discharge Methods

Approaches to brine waste treatment can vary depending not only on the contents on the stream, but also on the discharge or disposal method to be used. Brine waste treatment needs may range from minimal to intensive depending upon the chosen means of disposal, such as:

- Sewer discharge: Discharging brine waste to a publicly owned treatment works (POTW), or sewer, is generally a low-cost disposal option, but access to the POTW depends upon a facility's location. Additionally, the brine stream may need to be treated to suit the discharge standards of the POTW. Depending on the volume and the size of the receiving POTW, sufficient dilution of the brine waste stream may be achieved as it is combined with all the other streams entering the POTW.

- Surface water discharge: Brines can be discharged to lakes, rivers, or other bodies of surface water. While a relatively low-cost solution, facilities must typically comply with regulatory guidelines when discharging to waterways, which can mean adopting treatment strategies to remove certain contaminants or adjust the salinity, sometimes by dilution. Depending on the volume and the size of the receiving body of water, it is possible that a brine waste stream may be satisfactorily diluted by all the other streams entering it. If allowed by regulatory agencies, such disposal arrangements may mean minimal/no brine treatment is necessary.

- Evaporation: Evaporation ponds collect spent brine and leverage solar energy to remove water content through heat exposure. They require minimal investment in terms of operations, maintenance, energy, and compliance, but can take up a large footprint. New technologies, such wind-aided intensification of evaporation (WAIV), can reduce the footprint needed for a traditional evaporation pond and speed up the evaporation process. Brine waste treatment may still be necessary if toxic contaminants are present or if there is a large volume of waste brine.

- Recycling and reuse: Brines are commonly recycled for additional use, including industrial cooling, acid and caustic production, and many other applications. Increasingly, industrial plants are partnering with municipalities and other external entities to recycle brine waste in innovative ways, such as leveraging cheese brine for road deicing or using brine waste for irrigation of salttolerant land and vegetation. Brine waste streams often need to be treated prior to reuse, though the extent of treatment needed may be minimal (e.g. removing or reducing organic contaminants).

- Deep well injection: Deep well injection is typically reserved for applications requiring disposal of high-volume brine waste streams, such as those produced by hydraulic fracturing operations. This controversial disposal method involves pumping spent brine into porous rock formations found deep underground and is not permitted in all states. Deep well injection requires brine waste treatment for total suspended solids (TSS), oils, and greases, so as to prevent blockages in receiving wells. Additionally, deep well injection entails substantial costs associated with drilling, permitting, and monitoring the well, making it an impractical disposal option except for certain high-volume, highsalinity waste streams.

In some locations, costs for liquid waste disposal can run quite high. If this is the case, it may be worth considering zero liquid discharge (ZLD) systems, which remove all liquid components to minimize discharge volumes so as to cut costs. ZLD systems for brine waste treatment typically include a brine concentrator and crystallizer. While ZLD systems have relatively high capital and energy costs, they can more than make up for it by minimizing liquid waste disposal costs, as well as by facilitating reclamation of valuable materials.

Environmental Impact and Regulatory Compliance

Waste brines pose risks to the environment due to their high concentrations of salts, which can harm aquatic life when improperly discharged to the sea or surface waterways and can cause significant damage to soil and vegetation on land as well. Additionally, waste brines can contain toxic chemicals that may be harmful to humans and the environment. For this reason, disposal of waste brine is often strictly regulated and can be costly where large volumes are concerned.

It is always best to verify discharge, monitoring, and reporting requirements at your specific location to ensure that a chosen brine waste treatment solution will help your facility to stay in compliance with National Pollutant Discharge Elimination System (NPDES) permit standards or other discharge requirements.

Recycling Brine Waste for Reuse

Many industrial processes require brine in part of their process, such as hydrometallurgy, sodium hypochlorite, lithium carbonate, and chloralkali manufacturing plants, to name a few. Some facilities even use leftover brine for irrigation or deicing.

Regardless what your facility is using the recycled brine waste for, keep in mind that brine leftover from production often accumulates contaminants along the way, such as:

- Silica.
- Heavy metals.
- Hardness.
- Organic compounds.
- Sulfates, nitrates, and phosphates.
- Suspended solids.

The brine waste can be pretreated with coagulants, polymers, additives, and pH adjustment to settle out many of the larger contaminants, including metals, sulfates, and other suspended solids that can foul membranes and cells down the production line. Other contaminants, such as calcium, can scale equipment, so depending on the manufacturing process and brine requirements, there are several membrane and ion exchange technologies that can prevent these issues and produce a useful brine stream adequate for your process.

For example, if a chlor-alkali plant is looking to reuse brine waste that is contaminated with metals like iron, vanadium, or manganese, they will often have trouble with the fouling of downstream equipment, which can often lead to unscheduled plant downtime and other delays. The brine stream might be treated with specialty chelating resins that target specific metals for removal without being exhausted by sodium, thereby removing the unwanted contaminants while preserving the salt concentration in the brine solution.

When it comes to facilities that produce lithium carbonate or lithium hydroxide, specialty ion exchange resins can be used to remove unwanted metals and other contaminants, and the brine stream can then be further treated with high-pressure membranes that can separate out and concentrate the lithium carbonate and lithium hydroxide, so the technologies and methods of purification and separation vary greatly depending on the quality of the brine waste and the brine composition required for production.

Treating Brine Waste for Discharge

Most brines are technically considered nontoxic, but industrial brine waste released in heavy concentrations to the environment or to local publicly owned treatment works (POTW) can cause an assortment of issues if untreated. For example, if the waste contains more than just sodium, it can possibly harm aquatic life in local waterways. In some cases, simply diluting the brine waste prior to discharging might be a practical solution. It can also be against discharge regulations, which can cause your facility to pay a heavy fine.

Most discharge regulations will require a reduction in the amount of sodium, total suspended solids, and contaminants in general, which can often be treated with membrane technology such as ultrafiltration followed by reverse osmosis. Because of the high concentration of salt in brine, pretreatment is often used to help protect downstream filtration units and equipment.

Another option for treating brine waste is evaporation. This can be done in an outdoor evaporation pond or with a technique called vacuum evaporation. By evaporating the effluent under vacuum, the boiling point is reduced (thereby saving energy) and what's left is a crystalized mass of salt and, separately, a purified water stream. This method can be useful for drying brine filtrates thoroughly, and it can be implemented in tandem with other treatments to boost effectiveness.

Whichever method your facility uses to treat its brine waste for discharge, make sure it is done so in compliance with the facility's POTW or National Pollutant Discharge Elimination System permits.

Brine Treatment Technologies

Brine waste is typically either recycled for use in the facility's process or treated for disposal, but these processes are rarely straightforward or practical, and can even be costly.

A typical brine treatment system might include;

Membrane Filtration

One of the more cost-effective technologies for treating brine waste, membrane filtration is still widely used across various industries. When treating brine, it's common to see systems that begin with ultrafiltration (UF) and end with reverse osmosis (RO).

When UF is used prior to RO, it effectively removes various contaminants while protecting downstream membranes from premature wear and fouling. UF membranes are available in pore sizes ranging from 0.001 to 0.1 µm, which means UF removes a more comprehensive range of contaminants than some other membranes (such as microfiltration) while leaving behind ions and organic compounds of low molecular weight. UF is suited for removal of very fine particles, including proteins, colloidal silica, and silt.

The stream is then passed through an RO system, which is a membrane technology that uses a semipermeable medium to remove certain ions and particles from a liquid stream that are 0.0001 µm or larger, including salt. This produces a high-quality water stream in addition to a highly concentrated stream of waste, which can be used or discarded, depending on the needs of the facility.

This sequence of membrane treatment is effective for protecting equipment and minimizing chemical costs and system downtime for cleaning and is generally considered a low energy–consuming technology.

Evaporation and Crystallization

After brine is concentrated by membrane filtration, thermal processes or evaporation are often used as the next step to dry solids. Excess water is evaporated off, collected, and reused (adding acid at this point will help to neutralize the solution so, when heating it, you can avoid scaling and harming the heat exchangers). Deaeration is often used at this phase, as well, to release dissolved oxygen, carbon dioxide, and other noncondensible gases to further protect equipment from corrosion and other harmful occurrences.

The leftover waste then goes from an evaporator to a crystallizer, which continues to boil off all the water until all the impurities in the water crystallize and are filtered out as a solid. This process is often used in facilities that aim for zero liquid discharge, but it's generally reserved for only facilities that require it (most likely due to stringent environmental and discharge regulations) as it is considered a costly and high energy–consuming process.

Ion Exchange

Ion exchange (IX) systems are used across a variety of industries for water softening, purification, and separation purposes. These systems separate ionic contaminants from solution through a physical-chemical process where undesirable ions are replaced by other ions of the same electrical charge. This reaction occurs in an IX column or vessel where a process or waste stream is passed through a specialized resin that facilitates the exchange of ions.

When contaminant removal needs are highly specific, many times IX is ideal. Chelating resins are the most common type of specialty resin and are often used for brine softening.

Weak acid cation (WAC) exchange resins are also often used to remove cations associated with alkalinity (temporary hardness) and are therefore also ideal for softening brine. Facilities can also use various proprietary resins to separate contaminants, such as metals, from the brine stream to support production and manufacturing of lithium carbonate and lithium hydroxide.

Electrodialysis is also a form of ion exchange that can be used in the brine treatment system. It's a membrane process that uses positively or negatively charged ions to allow charged particles to flow through a semipermeable membrane and can be used in stages to concentrate the brine. It is often used in conjunction with RO to yield extremely high recovery rates. Combined, these technologies can concentrate a brine stream down to a high salinity while pulling out up to 60–80% of the water.

Other Things to Consider when Treating Brine

As mentioned previously, when looking at basic separation for a contaminated brine stream, you will always be left, essentially, with a more purified brine (such as a sodium chloride stream). The cost to treat a brine stream like this can be relatively low and would include treatments such as:

- Membrane filtration.
- Precipitation.
- Carbon adsorption of organics.
- Oil and water separation.

Using IX, the facility can take out easy-to-remove contaminants, which might leave you with something like a sodium chloride or brine sulfate brine stream, which is a salt stream that doesn't easily precipitate.

These technologies are relatively low cost compared to removing the salt directly out of the brine (such as with sodium chloride or sodium nitrate). In these instances, it's very difficult to remove the salts and almost always involve high-pressure membrane systems to concentrate the brines (pulling the water out of the brine).

In situations where membrane filtration won't be effective, evaporation and crystallization will help remove the water from the brine, but such thermal polishing processes are exponentially more expensive when it comes to purchasing the relative technology in addition to high operational and energy usage costs.

Essentially, by trying to solve one problem your facility can easily create another.

The only true ways to dispose of brine are to reuse it or dilute and slowly release it back into the environment. For example, if you have a very large river near your facility that is available of discharging your treated brine, you can slowly bleed the brine in, dissolving it back into the body of water with a minimal impact to the river. If you have a smaller stream, your facility will not be able to do this; doing so could pose catastrophic harm to the environment, killing off fish and other wildlife in the surrounding areas. In some areas, where permitted, waste brines can be deep-well injected although this option is becoming more restricted.

Facilities can reuse brine, but recycling it will concentrate it more, which would leave you discharging a smaller stream with a higher concentrate of salt and TDS.

As always, it is usually best to discuss any treatment options with a water-treatment professional who is able to analyze your facility's needs, specifically, and help you determine the best treatment options, as they can be highly individual.

While used in a number of industrial processes, brines can be difficult to treat and discharge once they've reached the end of their useable life. If your plant is looking for more economical ways to manage brine waste, you might be wondering "How can a facility reuse brine waste?"

Today there are more instances than ever of facilities successfully reusing brine waste, however, there is still a relatively narrow set of circumstances where brines can be effectively reused.

Why Recycle Brine Waste?

There are a number of brine waste disposal options, though many of them are controlled by permits and environmental regulations. Additionally, recent public attention to issues of water scarcity and environmental responsibility have spurred increasing public and regulatory scrutiny of popular brine disposal methods, including discharge to bodies of water as well as deep-well injection. Where discharge costs and limitations are concerned, brine waste reuse and recycling strategies can provide several benefits, such as reducing wastewater discharge volumes or allowing facilities to increase production volumes while staying within discharge permit limitations. Additionally, some facilities have seen a net cost savings by supplementing fresh water demand with waste brine, especially in regions where access to source water may be costly and unreliable.

Types of Brine Recycling and Reuse

The feasibility and cost of brine reuse and recycling can depend on a number of factors, including facility location, brine constituents, and volume. Commissioning a treatability study is one way to get a better understanding of the salinity levels, as well as metals, total dissolved solids (TDS), and other contaminants present in your waste brine so you can better determine whether brine waste treatment for recycling and reuse is a cost-effective option.

With appropriate treatment, waste brines can be used for a number of different applications. Below, we've summarized a few of the more typical brine reuse and recycling strategies:

Acid and Caustic Production

Caustic soda (sodium hydroxide) and chlorine are used extensively across industries ranging from oil and petroleum refineries to pulp and paper manufacturers. Production of these chemicals involves electrolysis of a sodium chloride brine solution, a process which involves consumption of large volumes of brine waste as well as large quantities of energy. To improve overall efficiency, acid and caustic producers have implemented brine electrolysis recovery strategies, allowing for reuse of brine waste through multiple electrolysis cycles. In other cases, mining operations and other types of facilities have successfully leveraged their brine byproducts by repurposing them for acid and caustic production. In this way, some facilities are able to consume a portion of their brine waste streams, while simultaneously producing chlor-alkali products that can be used or sold.

Softener Brine Recycling

Brines resulting from both ion exchange (IX) softeners and reverse osmosis (RO) systems can be treated and reused for additional IX softening cycles, thereby reducing overall water and salt consumption. Brine waste reclamation setups capture a portion of used brine, treat it (if necessary), and return it to the brine tank during IX resin regeneration. While brine recycling can be a great way to improve softener efficiency, it is generally viable only for softeners that use a high salt dose and that have a relatively high sodium chloride concentration following an active cycle. Additionally, brine reuse can result in an even more concentrated effluent that can be more difficult or costly to discharge.

Irrigation and Agriculture

In some cases, waste brines can be applied to land for irrigation and agricultural purposes. Generally speaking, waste brines used for agriculture must have relatively low TDS and be free of toxic constituents, such as heavy metals. In some regions, reuse strategies have included application of waste brines to salt-tolerant plants. Alternatively, some facilities have been able to leverage waste brines to produce liquid fertilizer, especially for brine waste streams that contain a mix of sodium and hardness. In this case, the brine is typically treated through an IX process to selectively remove sodium, while retaining hardness that is valuable to plant development.

Deicing and Dust Control

In an effort to cut discharge costs, some facilities have begun offering waste brines to cities and municipalities for treating roads threatened by ice, snow, and dust. Circumstances vary—meaning that different regions or municipalities may be more or less willing to accept waste brines for road treatment—because certain constituents, such as naturally occurring radioactive materials and chemicals from fracking or other processes, can be present and potentially harmful to the local environment or water supply. In some cases, however, local governments are willing to pay facilities for their waste brines, such as liquid waste brines from cheese production, mining, gas production, and other industries for use in road-treatment applications.

Recovery and Sale of Solid Salts

Solid salts can be recovered from brines through zero liquid discharge (ZLD) systems. These typically consist of a brine concentration step, followed by crystallization, and a solids dewatering step. While ZLD technologies can be costly, they are effective in eliminating wastewater discharge costs, and can be leveraged to produce solid salts. Brine byproduct salts can be used or sold for a variety of industrial applications, such as detergent manufacturing, dyeing, chemical production, curing, water treatment, and deicing, among other uses.

Cooling

While commonly used by power plants, oil refineries, chemical plants, and other facilities, cooling towers often consume large quantities of water. With proper treatment, recycled brine can be a great solution for reducing water use in cooling and thermal equipment. Recycling brine for this purpose will generally demand some form of cooling tower water treatment to minimize corrosion and scaling, which can include reduction of salt content, chlorides, phosphates, sulphates, and oxidizing bacteria. Recycling for cooling applications is limited by the salt concentration and other contaminates, as it inhibits heat transfer and may cause foaming.

PAPER MILL SLUDGE MANAGEMENT

The management of wastes, in particular of industrial waste, in an economically and environmentally acceptable manner is one of the most critical issues facing modern industry, mainly due to the increased difficulties in properly locating disposal works and complying with even more stringent environmental quality requirements imposed by legislation. In addition, in recent years the need to achieve sustainable strategies has become of greater concern, also because some traditional disposal options, such as landfill, are progressively restricted, and in some cases banned, by legislation. Therefore, the development of innovative systems to maximize recovery of useful materials and energy in a sustainable way has become necessary. Industrial wastes are generated through different industrial processes or energy production utilities as additional materials. Industrial symbiosis theory defines non-deliberately produced material as by-products or valuable raw materials which can be exploited in other industrial avenues. Paper industry is a strategic industry in many countries but in the same time, the production of paper consumes high quantities of energy, chemicals and wood pulp. Consequently, the paper production industry produces high environmental emission levels mainly as CO_2 due to energy consumption, or solid waste streams which include wastewater treatment sludges, lime mud, lime slaker grits, green liquor dregs, boiler and furnace ash, scrubber sludges and wood processing residuals. In terms of volume, most solids or liquids are those from the treatment of effluents, although waste from wood is also produced in large quantities.

Trends in Paper Mill Sludge Generation

The pulp and paper industry produce over 304 million tons of paper per year. In 2005, soley in Europe, 99.3 million tons of paper was produced which generated 11 million tons of waste and represented 11 % residue in relationship to the total paper production. The production of recycled paper, during the same period, was 47.3 million tons, generating 7.7 million tons of solid waste which represented 16 % of the total production from this raw material. Global production in the pulp and paper industry is expected to increase by 77 % by 2020 and over 66 % of paper will be recycled at the same time. On average, the majority of waste generated from paper production and recycling is PMS, which is a by-product of up to 23.4 % per a unit of produced paper, the quantity depending on paper production process. The countries joined in the CEPI organization itself produce more than 4.7 million tons of PMS per year and global production of PMS was predicted to rise over the next 50 years by between 48 and 86 % above the current level. This represents an enormous environmental burden due as more than 69 % of the generated PMS is landfill disposed. Flowchart of PMS generation is shown in figure.

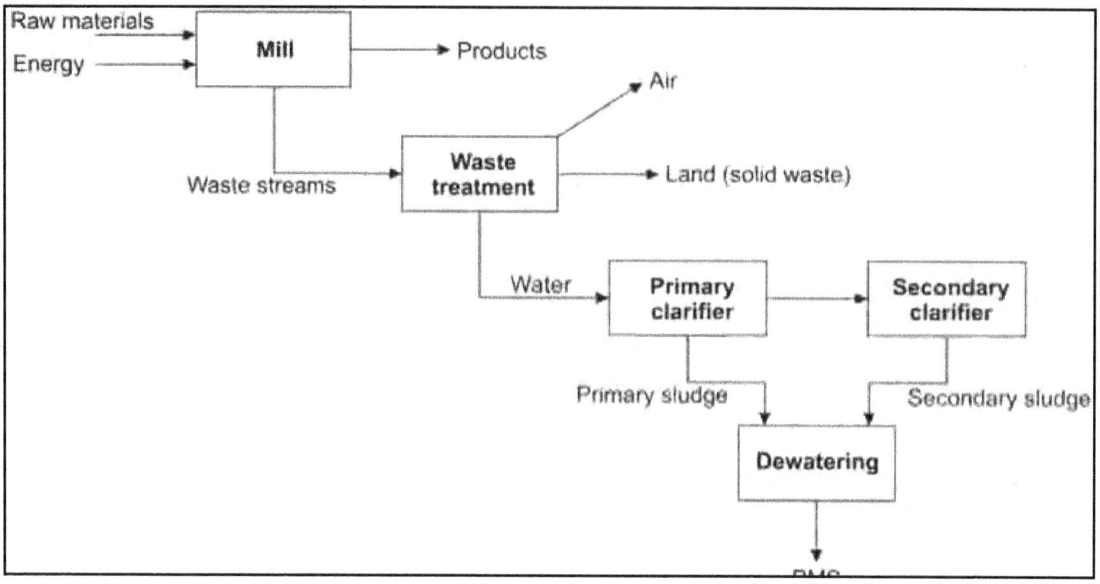

Scheme of PMS generation.

Table: Composition, heating valucs and cation exchange characteristics (CEC) for different PMS origins.

	PMS origin	Organic content wt %	Ash content wt %	Heat value MJ/kg	CEC cmol/kg
Low-ash sludges	Primary sludge from pulp mill producing pulp from virgin wood	94.31	5,69	20,1	30,20
	Primary sludge from paper mill producing paper from virgin wood	93,79	6,21	19,8	32,41
High-ash sludges	Secondary sludge from paper mill producing paper fom recycled cellulose without deinking process	67,23	32,77	16,5	17,30
	Primary sludge from paper mill producing paper from recycled cellulose without deinking process	64,72	35,28	14,2	33,58
	Deinking sludge from papermill that recycled paper not recycled previously	60,36	39,64	12	18,12
	Deinking sludge from papermill producing newspaper	59,30	40,70	12,2	19,06

Recyclers are producing two to four times more sludge as Virgin pulp mills but the characteristics of PMS vary, depending on pulp and paper mill processes and from raw materials entering into those processes. Produced sludge can be considered to fall into two main types: high-ash sludge (> 30% dry weight) and low-ash sludge (< 30% dry weight). High-ash sludges are chemical flocculation sludges generated by pulp mills, primary sludges generated by production of paper from recycled fibers and deinking sludges generated by paper mills, alternatively, low-ash sludge represents primary, secondary or biological sludges generated by pulp or paper mills.

Primary and deinking PMS constitutes a mixture of short cellulosic fibers and inorganic fillers, such as calcium carbonate, china clay, and residual chemicals dissolved in the water.

SEM image show complex structure of PMS (cellulose fibers and china clay).

Currently, PMS are considered to be a waste material. Increased recycling over the past two decades has consequently increased the amount of material that requires disposal. The paper industry implements several methods to dispose of the sludge that pulp and paper production generates.

Major Concerns with Paper Mill Sludge Management

Landfilling

PMS with high organic content in landfill is subjected to aerobic and anaerobic decay. According to Buswell and Mueller, 1 ton of low-ash PMS in landfill theoretically releases into environment approximately 2.69 tons of CO_2 and 0.24 ton of CH_4. Currently, most PMS is dried, spread or deposited onto the landfill. The landfills can be industrial, in that are constructed and operated by the mills, or they can be independently owned, requiring the mills to pay a "tipping fee" for sludge disposal. The European Landfill Directive and upcoming bio-waste directive aims to prevent or reduce, as far as possible, the negative effects of waste landfill on the environment, by introducing stringent technical requirements for wastes and landfill. By increasing landfill fees up to 3.5 €/ton annually until the final landfill use fee will reach 40 €/ton. That fact and adaptation of paper mills to Integrated Pollution Prevention and Control Directive 1996/61/CE forced the paper mill industry to look for processes for their waste minimization as well new technologies for waste reuse or environmental friendly disposal.

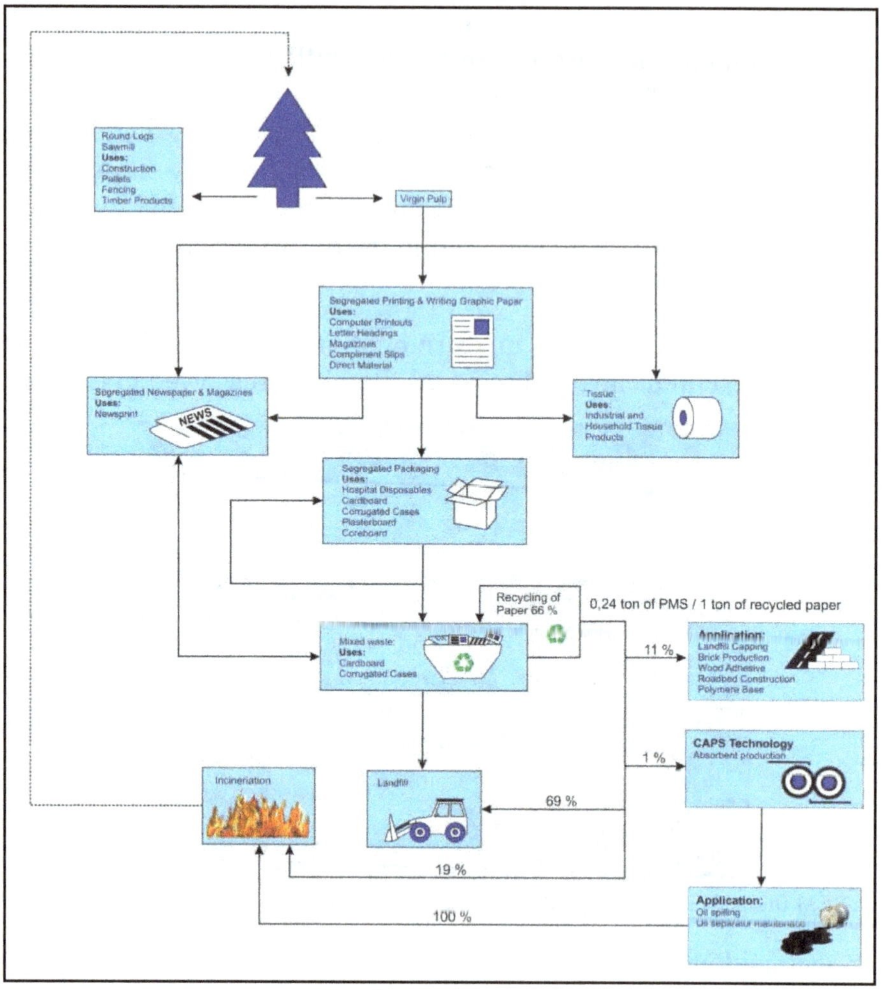

PMS disposal pathways.

Landspreading

Land surface spreading is one of the possible methods for the application of sludge to forest or agricultural lands for soil enhancement. PMS from pulp mills and secondary sludges are not hazardous due to heavy metals. On the other hand, sludge from deinking and waste paper mills may be relatively high in heavy metals due to the formulations used in ink removal. Compounds that can be found in mill sludges at concentrations above 10 mg/kg of dry sludge include naphthalene, phthalates, chloroform, PCBs, wood extractives or derivates and chlorinated lignin derivates. Sludge as land spreading material must be used very carefully due to an unfavorable C:N ratio, high ion exchange capacity and possible Cr toxicity. With maximum application rates of 12 tons per 405 ha, even a moderately sized mill will need a large area to spread its sludge. The use of land spreading material adjacent to residential areas is questionable due to odor.

Composting

Composting is one of stabilization techniques used for prevention of uncontrolled PMS decay. The use of compost is questionable due to the same reasons mentioned in land spreading. The C:N ratio in the sludge is up to 930:1 and it is not appropriate for plant growth. Composting of PMS requires a large land area and additional costs for storage and processing.

PMS for Energy

In practice approx. 19 % of PMS is incinerated on the sites due to energy recovery, but the economics of incineration is questionable because PMS contains 30 to 50 % water and only 30 to 50 % of cellulosic fibers calculated on dry solids. For each additional 1 % of moisture content in PMS, the temperature of combustion must be 10 °C higher due to process efficiency. The European Waste Incineration Directive 2000/76/EC intends to prevent or reduce, as much as possible, air, water and soil pollution caused by the incineration or co-incineration of waste, reducing at the same time the risk to human health that incineration processes entail. However, under controlled conditions PMS can become a sustainable fuel for co-generation.

Many technologies to utilize different PMS for energy exploitation have been tested and used. That processes include efficient incineration in fluidized bed and circulating fluid bed combustion chambers as well as more sophisticated processes for energy recovery as pyrolysis, indirect steam gasification, wet oxidation, super critical water oxidation and gasification.

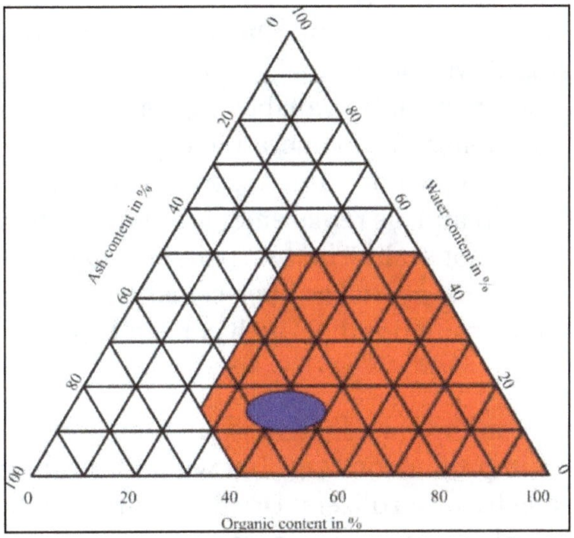

Fuel triangle for waste from paper industry (blue spot – deinking PMS).

Utilization in Brick, Light Aggregates and Cement Production

PMS containing a high inorganic fraction can be utilized in the production of building materials. Due to its combustion matrix, it can be used in the brick production industry. The addition of 5-15 % of PMS as raw materials improves both the final product and the processes. First, since its fiber content increases the porosity of the matrix, it enables the manufacture of lighter bricks; second, it saves fuel in the oven, decreasing firing time and makes the product more resistant against cracking during the drying and firing stages. The same advantages can be used in the production of light aggregates for the building industry. A similar exploitation has been noticed in cement industry. PMS with high organic content has an energy level that makes it an efficient alternative fuel in the production of Portland cement. Currently it is classified as Class 2 (liquid alternative fuels) in the Cembureau classification of alternative fuels. Including up to 20 % of deinking sludge into mortar improves the mortar mechanical properties.

Landfill Capping Material

Significant progress has been made in the use of paper mill sludge as a material for land fill cover by replacing the clays or geo-composites. PMS behaves similar to a highly organic soil and has good chemical, hydrodynamic and geotechnical properties which make it an efficient impermeable hydrodynamic barrier for the land field cover.

Combining the paper mill sludge as hydrodynamic impermeable barrier with metal scoria as an oxidizing layer an efficient landfill capping system can be build. More than 21 different studies have been carried out in 2009 on the use of PMS in the landfill body. The characteristics of PMS have been exploited also for road bed construction for light loaded roads and tennis courts.

Advanced Research and Utilisation of PMS

According to modern industrial trends on eco symbiosis economic efficiency is key factor for planning modern technological processes. This includes decreasing the waste streams during production and the use of produced waste as by-products or raw materials with higher added values. As can be seen from Figure PMS is a chemically and physically complex material which can be used in different industrial applications. One of the promising innovations is the use of PMS as heat insulation material with a thermal conductivity factor lower than 0.055 W/m^2K. This is comparable to insulation materials available on the market currently. The US company, Minergy, has implemented technology for incineration of PMS under controlled conditions. The product of incineration is a highly efficient water absorbent composed of meta-kaoline granules. Using the PMS, generated by pulp mill, as replacement for virgin fibers in production of fiber-cement sheets is already tested in pilot scale. The cost benefits analysis showed environmental and economic benefits for PMS generator as well as fiber-cement sheets producer. Using the PMS for production of construction and insulation boards has been topic of The Waste and Resources Action Programme.

The use of the PMS as a paper and wood adhesive is an interesting concept under investigation. The use of PMS as cat litter is in production stage but without serious market demands. PMS can be dried in mixture with pesticides or fertilizers. Dried mixtures can be used as a pesticide or fertilizer carrier in agriculture which allows more controlled releasing active substances into the soil. Contamination associated with the sludge would be the ongoing threat in such a disposal route. Conversion of the lignocellulosic portion of PMS into methyltetrahydrofuran, which can be mixed with ethanol and natural gas into cleaner burning fuel, is already developed and has been found economically viable. Bio-conversion of PMS into ethanol is less viable due to toxic impurities in deinking sludges.

Utilisation PMS for Sorbent Production

Promising research has been conducted to use PMS as an oil sorbent material and use of the PMS as sorbent material is well documented, but currently the market was non receptive to such sorbent material due to cheap and efficient synthetic absorption material. The results of research studies have shown that PMS can be indirectly used as an active absorbent by converting it into activated carbon. It can be used as binding material for the removal of heavy metals ions from water, removal of phenols and as an absorbent for hard surfaces cleaning. A variety of the processes and different absorbent products have been developed for commercial purposes. One of the processes which have been developed for the production of a floor absorbent, in the form of a granular product is known as the KAOFIN process and it is described in U.S.

In U.S. Patent 4374794 the sludge is evaporated, extruded into pellets and dried at temperatures ranging from 100 °C to 150 °C, in order to form an oil absorbent material. However, modern industry faces frequent and serious oil spills and subsequent sanitation demands high costs for sorbent materials. Offering a cheap and efficient natural material such PMS could become a welcome solution. The CAPS (Conversion of paper mill sludge into absorbent) is an eco-innovation solution in the 'market uptake' phase, therefore prior to expanding industrialization in Slovenia and Finland. The CAPS process uses the surpluses of the thermal energy which paper mills usually waste into the environment for sorbent production. In addition, CAPS uses paper mill waste as a secondary raw material and converts it into a high added value absorbent. The technology is relatively cheap, simple and easily replicable particularly in markets with a developed paper industry. It is based on drying of PMS to the point where it can be efficiently mechanically and chemically treated to release cellulosic fibers from its inorganic matrix. The humidity of the deinking and primary paper mill sludge lies between 50 to 70 %, whereas the content of cellulosic fibers is approx. 52 %, the remainder is inorganic. After drying between 70 - 80 % of the solid content, PMS proceeds through special mechanical treatment (unraveling). This stage is crucial for the entire process due to the fact that in this section cellulose fibers are released from the inorganic matrix, which in turn allows material to float on the fluid surface.

However, the mechanical treatment expanded the surface area, but the sorbency was not linear with regards to the surface area.

Table: Particle size, active surface area and sorbency of used mineral oil versus mechanical treatment of PMS.

	Avg.particle size mm	Active surface area m^2/g	Sorptivity g/g
Untreated PMS	10	4.8096	1.23
PMS Grinded	4.18	3.2048	2.07
PMS Unraveled	1.67	36.0526	4.4
PMS Fluffed	0.7	2.9626	7.12

Raw PMS (left), unraveled PMS (middle) and fluffed PMS (right).

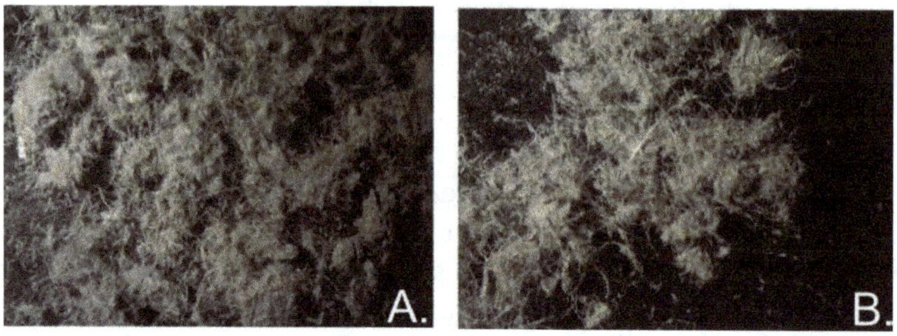

Microscope pictures of raw PMS (A) and unraveled PMS (B) at 8 x magnification.

As can be seen from figure the same trend was observed in sorbency of different substrates.

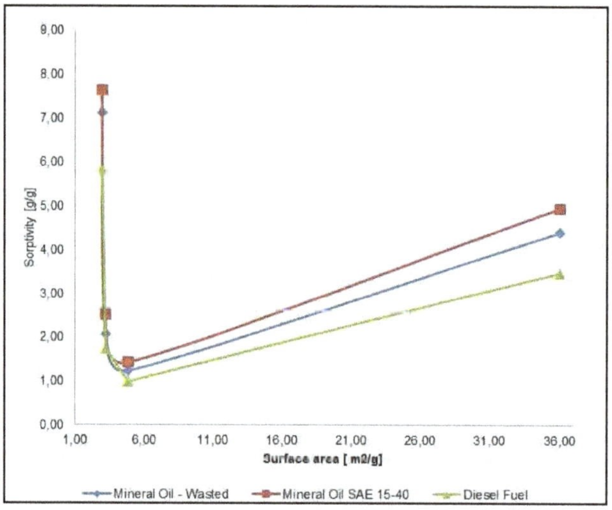

Active surface area of PMS absorbent samples vs. mechanical treatment.

The mechanical treatment was clearly connected to the breaking of the inorganic matrix into dust. The dust dropped off from the absorbent, which led to the shrinking of the surface area but made the remainder of the surface more accessible to the substrate. More violent mechanical treatment during the production (fluffing) of the absorbent led the isothermal sorption of the absorbent to become similar to the isothermal sorption of the paper standard. The results showed that the appropriate treatment process (especially the mechanical treatment phase) was of the utmost importance for the conversion of the PMS into a sorbent. The possible explanation for this appears to be that the PMS consisted of a fragile net structure composed of cellulosic fibers and an inorganic core, which served as an anchor for linking the cellulosic fibers together. The inorganic core played an important role in the sorption process because it served as absorption points for the hydrophobic substances, while the cellulosic fibers served as a floating skeleton. Chemical treatment (esterification, silanisation) is an option when higher absorption capacity and better buoyancy is required. Distributed (chemically prepared) paper mill sludge is dried at 130 °C to 150 °C until the humidity oscillates between 1% and 10% in order to obtain a final product with active surface area 36 m^2/g, sorption capacity up to 8 g oil/g PMS and capacity to float on the water surface.

Produced sorbent has a calorific value around 3.8 MJ/kg and when it is completely soaked by oil it has a calorific value up to 33.5 MJ/kg and can be used as high quality fuel in the cogeneration processes. Incineration under controlled conditions leads to the conversion of the kaolin portion of paper mill sludge into a meta-kaolin substance in the form of vitrified granules. These vitrified granules can be reused as an inert hydrophilic absorbent.

The adsorption kinetic shown, that the mechanism of PMS sorption follows complex combinations of interparticle diffusion at the first stage of the process followed by pseudo-second order adsorption of oil into pores of the inorganic part of PMS.

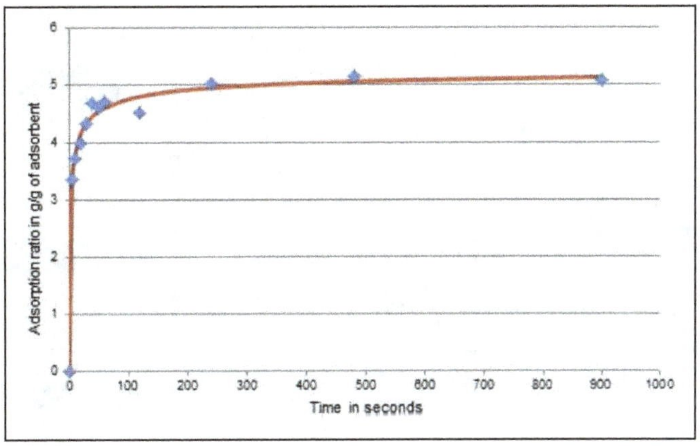

Sorption kinetics curve for mineral motor oil (SAE 15-40W) sorbing on PMS at 20°C.

In the loose state of adsorbent the adsorption equilibrium is reached after 240 seconds at 20°C and as can we see from the figure PMS absorbs more than 95 % of oil from a water surface within 70 seconds.

Environmental Impact Assessment of PMS Sorbent versus Expanded Polypropylene Sorbents

LCA model based on figure showed that conversion of PMS from paper production can expand paper life cycle for additional step and helps closing the life cycle in paper production processes. A model presenting the life cycle assessments (LCA) of PMS sorbent versus expanded polypropylene sorbents, is a basis for discussion about the sustainability of conversion of industrial waste from the paper industry into value added products. LCA of PMS and expanded polypropylene (EPP) were accomplished by using the BUWAL 250 database included in the software package SimaPro 7.1, and data from peer reviewed literature.

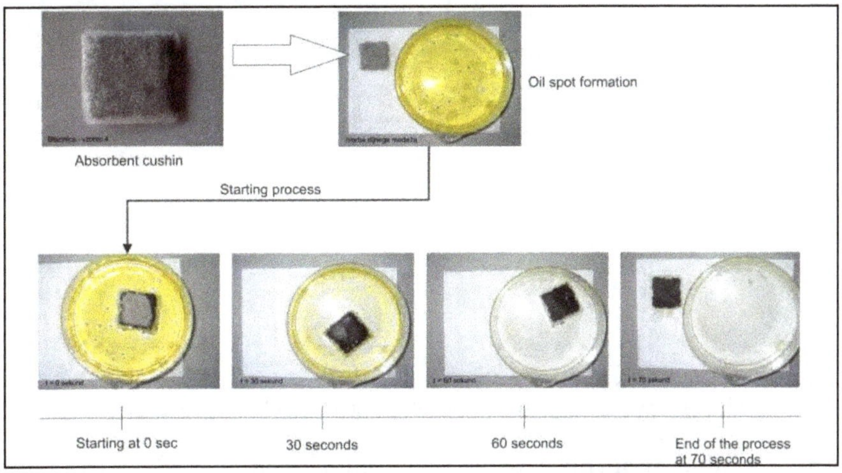

PMS sorption efficiency in the case of sorption of transformer oil from a water surface.

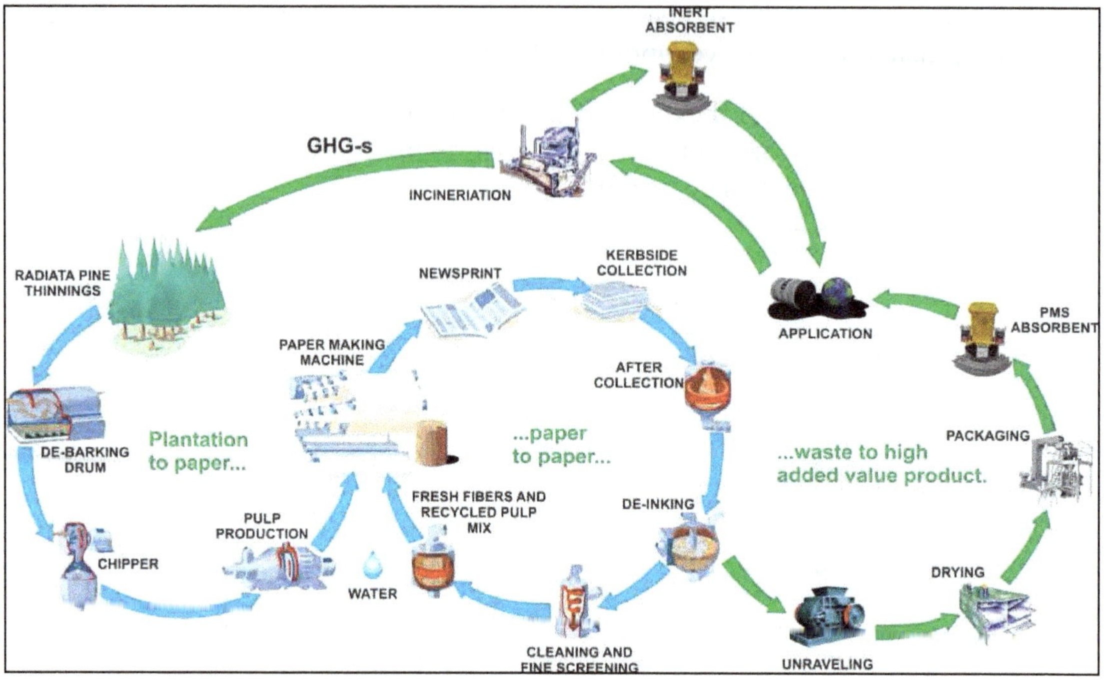

Life cycle of paper in the case of conversion of PMS into sorbent material.

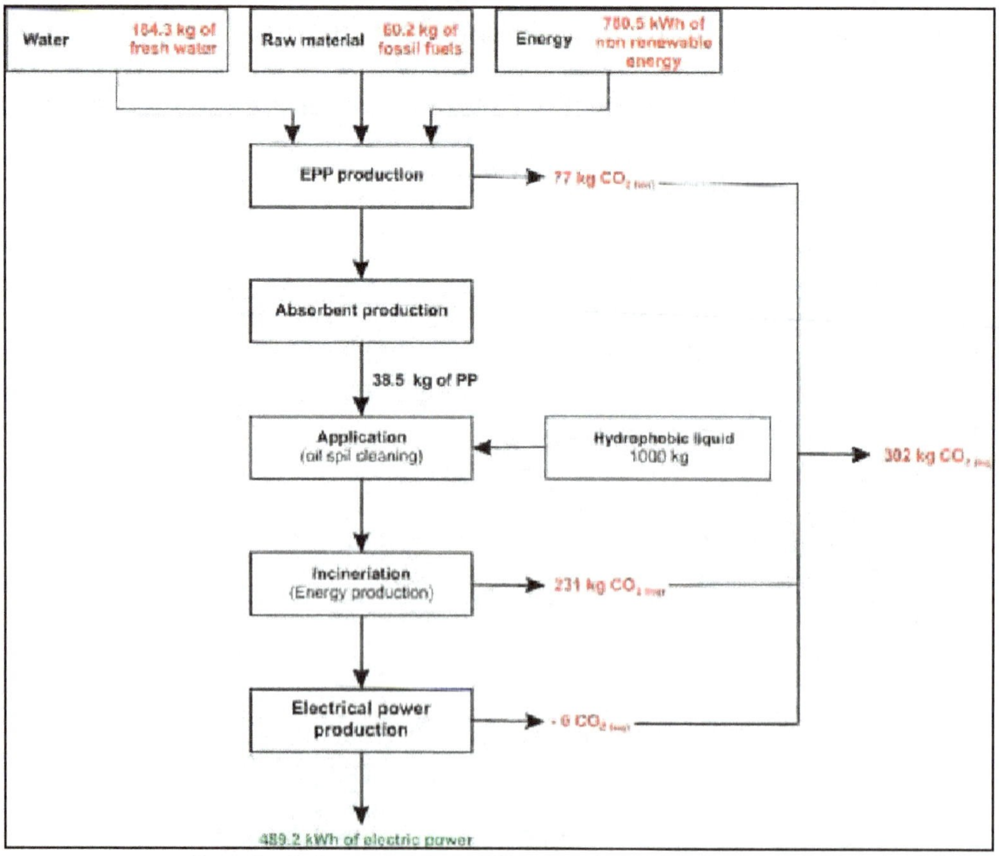

Life cycle circle production and application of EPP absorbent.

The data for the uncontrolled anaerobic decay were calculated according the equation for anaerobic decay and the data for energy production were calculated according to Martin et al. Paper mill sludge was considered as unwanted by-product in the paper and cardboard recycling processes that is without water and energy consumption and without GHG-s generation for its production. EPP was considered as absorbent produced from non-renewable raw materials (propene). The calculations were based on the quantity of absorbent required for the absorption of 1000 kg of oil with the following presumptions: the sorption capacity of absorbent produced from PMS and from EPP were 7 kg of oil/kg sorbent and 26 kg of oil/kg sorbent, respectively. It was considered that used sorbents were incinerated for energy recovery. Absorbed hydrophobic substances were not included in LCA calculations.

As can be seen from Figures, the production and application of PMS as a sorbent for oil spill sanitation reduced carbon footprint by 2.75 times when compared to the production and application of EPP. In addition, considering the production of absorbent instead of landfill disposal the PMS decreased carbon footprint by 5.25 times. Altogether, production of the sorbent material from PMS instead of landfill disposal it and replacing the synthetic EPP absorbent with PMS absorbent for oil spill sanitations reduced the carbon footprint by more than 14 times. The difference in the water balance was 372.3 kg calculated on the quantity of sorbent needed for sanitation of the 1000 kg of oil. The difference in the water balance was due to the consumption of the 184.3 kg of fresh water for EPP production and due to the production of 143 kg of clean technological water in the process of energy recuperation during PMS sorbent production. The energy balance showed the surplus of energy in the production of the PMS absorbent was due to the production of a combustible product from waste and negative energy balance in the case of the production and use of EPP absorbent. The later was due to heavy energy consumption during production of PP from fossil fuels. Additionally, the LCA analysis has shown that conversion of PMS into an absorbent prolongs life cycle of paper products for an additional two cycles and efficiently closes the life cycle circle of paper.

The modern sustainable management of production processes should be based on the industrial ecology approach, of which an essential element is the eco-symbiosis theory. Pulp and paper industry producing enormous quantities of solid waste what presents huge environmental burden. Appropriate managing with such a waste is most crucial task for modern pulp and paper industry. Many innovative approaches for conversion of the PMS into useful materials have been done in past two decades, but for many of them the markets demands have been too small for successful diverting of PMS from the landfill disposal. The CAPS process for the conversion of PMS into a sorbent for water surface cleaning fulfills all of above mentioned requirements. Moreover, it also includes the surpluses of thermal energy which paper mills usually transmit into the environment. The manufactured natural sorbent may be used by the oil, chemical, logistic and transport industries as well as public bodies such as fire brigades, civil protection and disaster relief institutions. These are institutions which all require an environmental friendly, efficient, cheap and at the same time sustainable product for cleaning of oil spills from water surfaces and for oil separators maintenance and on the other hand the pulp and paper industry will expand their product portfolio by using waste as raw material. The main drawback of placing the PMS sorbent on the market is its slow degradation in water. That obstacle can be overcome by chemical treatment, proper application and by mixing PMS sorbent with floating materials but those processes raise the production costs and makes PMS absorbent unattractive to the market. Additional improvement must be made in the future to overcome water degradation of the PMS sorbent and with success the industry can obtain a sustainable, cheap sorbent for oil spill sanitation in unlimited quantities.

References

- Understand-industrial-wastewater-treatment: chemicalprocessing.com, Retrieved 25 June, 2019
- Effects-wastewater-environment: organicawater.com, Retrieved 14 April, 2019
- Treatment-of-industrial-effluents-primary-secondary-and-tertiary, waste-management-1835: environmental-pollution.in, Retrieved 10 May, 2019
- Examples-for-bio-treatment-of-industrial-effluents, industrial-effluents- 3782: environmentalpollution.in, Retrieved 17 July, 2019
- Anaerobic-biological-wastewater-treatment: emis.vito.be, Retrieved 19 March, 2019
- Zerosludge-treatment-of-industrial-wastewater-by-ken-norcross, municipal-technologies- 16211488: water-world.com, Retrieved 28 August, 2019
- Marcus Van Sperling (2007). Activated Sludge and Aerobic Biofilm Reactors. IWA Publications. ISBN 1-84339-165-1

PERMISSIONS

All chapters in this book are published with permission under the Creative Commons Attribution Share Alike License or equivalent. Every chapter published in this book has been scrutinized by our experts. Their significance has been extensively debated. The topics covered herein carry significant information for a comprehensive understanding. They may even be implemented as practical applications or may be referred to as a beginning point for further studies.

We would like to thank the editorial team for lending their expertise to make the book truly unique. They have played a crucial role in the development of this book. Without their invaluable contributions this book wouldn't have been possible. They have made vital efforts to compile up to date information on the varied aspects of this subject to make this book a valuable addition to the collection of many professionals and students.

This book was conceptualized with the vision of imparting up-to-date and integrated information in this field. To ensure the same, a matchless editorial board was set up. Every individual on the board went through rigorous rounds of assessment to prove their worth. After which they invested a large part of their time researching and compiling the most relevant data for our readers.

The editorial board has been involved in producing this book since its inception. They have spent rigorous hours researching and exploring the diverse topics which have resulted in the successful publishing of this book. They have passed on their knowledge of decades through this book. To expedite this challenging task, the publisher supported the team at every step. A small team of assistant editors was also appointed to further simplify the editing procedure and attain best results for the readers.

Apart from the editorial board, the designing team has also invested a significant amount of their time in understanding the subject and creating the most relevant covers. They scrutinized every image to scout for the most suitable representation of the subject and create an appropriate cover for the book.

The publishing team has been an ardent support to the editorial, designing and production team. Their endless efforts to recruit the best for this project, has resulted in the accomplishment of this book. They are a veteran in the field of academics and their pool of knowledge is as vast as their experience in printing. Their expertise and guidance has proved useful at every step. Their uncompromising quality standards have made this book an exceptional effort. Their encouragement from time to time has been an inspiration for everyone.

The publisher and the editorial board hope that this book will prove to be a valuable piece of knowledge for students, practitioners and scholars across the globe.

INDEX